P9-BHU-112

CA. STATE UNIVERSITY, SACRAMENTO
3 0600 00668 3214

THE GREENING OF INDUSTRIAL ECOSYSTEMS

BRADEN R. ALLENBY and DEANNA J. RICHARDS, *Editors*

NATIONAL ACADEMY OF ENGINEERING

NATIONAL ACADEMY PRESS
Washington, D.C. 1994

NATIONAL ACADEMY PRESS • 2101 Constitution Ave., NW • Washington, DC 20418

The National Academy of Engineering was established in 1964, under the charter of the National Academy of Sciences, as a parallel organization of outstanding engineers. It is autonomous in its administration and in the selection of its members, sharing with the National Academy of Sciences the responsibility for advising the federal government. The National Academy of Engineering also sponsors engineering programs aimed at meeting national needs, encourages education and research, and recognizes the superior achievement of engineers. Dr. Robert M. White is president of the National Academy of Engineering.

This volume has been reviewed by a group other than the authors according to procedures approved by a National Academy of Engineering report review process. The interpretations and conclusions expressed in the symposium papers are those of the authors and are not presented as the views of the council, officers, or staff of the National Academy of Engineering.

Funding for the activity that led to this publication was provided by the W. M. Keck Foundation, the Andrew W. Mellon Foundation, and the National Academy of Engineering Technology Agenda Program.

Library of Congress Cataloging-in-Publication Data

The greening of industrial ecosystems / Braden R. Allenby and
Deanna J. Richards, editors.
p. cm.
Includes bibliographical references and index.
ISBN 0-309-04937-7
1. Environmental sciences. 2. Factory and trade waste.
3. Environmental policy. 4. Conservation of natural resources.
I. Allenby, Braden R. II. Richards, Deanna J.

GE105.G74 1994 93-46753
363.73'1—dc20 CIP

Cover art: *Spinozza*, mixed media on paper, courtesy of the artist, Karen Vogel, Washington, D.C.

This book is printed on recycled paper. 

Printed in the United States of America

Steering Committee Members

ROBERT A. FROSCH, Senior Research Fellow, John F. Kennedy School of Government, Harvard University, and Senior Fellow, National Academy of Engineering
ERNEST L. DAMAN, Chairman Emeritus, Foster Wheeler Development Corporation
SHELDON K. FRIEDLANDER, Parsons Professor of Chemical Engineering, and Director, Air Quality/Aerosol Technology Laboratory, University of California at Los Angeles
HENRY R. LINDEN, Max McGraw Professor of Energy and Power Engineering and Management, Illinois Institute of Technology
WILLIS S. WHITE, JR., Retired Chairman, American Electric Power Company, Inc.

Staff

DEANNA J. RICHARDS, Project Director
BRADEN R. ALLENBY, J. Herbert Hollomon Fellow (January–December 1992)
MARION R. ROBERTS, Senior Program Assistant

Preface

Industrial ecology is the study of the flows of materials and energy in industrial and consumer activities, of the effects of these flows on the environment, and of the influences of economic, political, regulatory, and social factors on the flow, use, and transformation of resources. The objective of industrial ecology is to understand better how we can integrate environmental concerns into our economic activities. This integration, an ongoing process, is necessary if we are to address current and future environmental concerns.

We have made great progress in the last two decades in targeting specific sources of pollutants and controlling them at the ends of pipes, the tops of smokestacks, and in landfills. In the 1990s, however, many of our concerns are on sweeping regional and global issues such as acid rain, stratospheric ozone depletion, global climate change, and dispersion of heavy metals throughout the biosphere. These new concerns and the approaches to addressing them require improved understanding of the environmental effects of industrial use and transformation of materials and energy, including consumer use and disposal of products. It also requires a better appreciation of the factors that influence such use, transformation, and disposal. This understanding can inform efforts to minimize environmental degradation.

At the most general level, the reshaping of industrial systems for environmental and economic success is based on efficient use of materials and energy, substitution of more abundant and environmentally preferable materials for those that are rare or environmentally problematic, reuse and recycling of products and materials, and control of waste and emissions. More detailed examination of the integration of environmental concerns into economic decision making raises a plethora of questions, beginning with the one that opens the overview of this book: "Paper or plastic?" This leads to more general questions: How do we assess the environmental preferability of one material over another? What improvements do we need in the design and management of materials or products? How can we integrate new environmental design and engineering concepts into industry practices and into engineering and management education? How can we overcome the failure to adopt existing technologies and methodologies that represent improvements over prevailing industrial practices? To what extent is that failure a result of competing priorities in management and marketing, or of conflicting economic incentives? How do we move to market-based initiatives for internalizing environmental costs when that generally implies increasing prices on less environmentally desirable materials, products, or practices?

This volume is a product of the National Academy of Engineering (NAE) continuing program on Technology and Environment. It examines the greening of industrial systems through the lens of industrial ecology. It examines promising approaches to environmentally conscious design and manufacturing, as well as education and research needs. It promotes greater recognition of environmental dimensions in formulating technology policies and management strategies in both the public and the private sectors.

The case studies of emerging business practices described in this volume suggest fundamental management strategies of corporate environmental stewardship. These are in keeping with world-class manufacturing practices such as commitment of senior management to change, the setting of goals and priorities, using cross-function teams, baselining, forming partnerships with stakeholders, and managing the supplier chain. A rich set of examples and concepts on the subject of industrial ecology and environmentally conscious design is presented in this publication.

This book is the result of a two-year effort, based on a workshop held at Woods Hole, Massachusetts, in July 1992. The workshop and planning effort were chaired by Robert Frosch. The concept for this activity may be traced to a proposal developed by Bruce Guile and Deanna Richards. In addition to these individuals, special thanks go to Braden Allenby, the NAE's second J. Herbert Hollomon Fellow, who organized the workshop, provided valuable ideas and contributions to the project, and in general helped further the NAE's effort in this area of industrial ecology. We are indebted to the workshop steering committee members (whose names appear on p. iii), to the authors for their excellent chapters, to an editorial team consisting of Braden Allenby, Deanna Richards, Dale Langford, Bette Janson, and Marion Roberts, and to Bruce Guile, director of the NAE Program Office, for his advice and assistance on the project and publication.

Finally, I would like to express my appreciation to the W. M. Keck Foundation for its generous support of this project and to the Andrew W. Mellon Foundation for supporting related elements of the NAE's program on Technology and Environment.

Robert M. White
President
National Academy of Engineering

Contents

The Greening of Industrial Ecosystems: Overview and Perspective 1
Deanna J. Richards, Braden R. Allenby, and Robert A. Frosch

UNDERSTANDING INDUSTRIAL ECOLOGY AND ITS CONTEXT

Industrial Metabolism: Theory and Policy 23
Robert U. Ayres

Energy and Industrial Ecology 38
Henry R. Linden

Input-Output Analysis and Industrial Ecology 61
Faye Duchin

Wastes as Raw Materials 69
David T. Allen and Nasrin Behmanesh

Economics and Sustainable Development 90
Pierre Crosson and Michael A. Toman

From Voluntary to Regulatory Pollution Prevention 98
Frederick R. Anderson

International Environmental Law and Industrial Ecology 108
Robert F. Housman

Industrial Ecology: The Role of Government 123
Matthew Weinberg, Gregory Eyring, Joe Raguso, and David Jensen

EMERGING INDUSTRIAL ENVIRONMENTAL PRACTICE

Integrating Environment and Technology: Design for Environment 137
Braden R. Allenby

Preventing Pollution and Seeking Environmentally Preferable Alternatives in the U.S. Air Force 149
Edward T. Morehouse, Jr.

Designing the Modern Automobile for Recycling 165
Richard L. Klimisch

Greening the Telephone: A Case Study 171
Janine C. Sekutowski

The Utilization-Focused Service Economy: Resource Efficiency and Product-Life Extension 178
Walter R. Stahel

Zero-Loss Environmental Accounting Systems 191
Rebecca Todd

Implications of Industrial Ecology for Firms 201
Patricia S. Dillon

Design for Environment: An R&D Manager's Perspective 208
Robert C. Pfahl, Jr.

EDUCATION AND RESEARCH NEEDS

The Two Faces of Technology: Changing Perspectives in Design for Environment 217
Sheldon K. Friedlander

Industrial Ecology and Design for Environment: The Role of Universities 228
John R. Ehrenfeld

Biographical Data 241

Index 251

THE GREENING OF INDUSTRIAL ECOSYSTEMS

The Greening of Industrial Ecosystems. 1994.
Pp. 1–19. Washington, DC:
National Academy Press.

The Greening of Industrial Ecosystems: Overview and Perspective

DEANNA J. RICHARDS, BRADEN R. ALLENBY, and ROBERT A. FROSCH

"Paper or plastic?" is rapidly becoming the commonest query of our day. In selecting a desirable bag at the grocery store check-out stand, the environmentally concerned base their decision not just on which is better at carrying groceries, but which is "greener." Making a rational choice on environmental preferability of even simple options, however, is not simple. A comparison of beverage cups—plastic vs. paper vs. ceramic—illustrates the difficulty. The intuitive choice, based on disposability, is that ceramic is preferable, followed at a considerable distance by paper and then perhaps plastic, which is less biodegradable than paper. The most comprehensive study to date indicates, however, that ceramic cups are the preferable option only if each cup is reused at least hundreds of times; a conclusion dependent heavily on assumptions about washing patterns (van Eijk et al., 1992). As multiple reuse is unlikely in many cases (think of the logo-stamped coffee mugs that populate trade fairs), the counterintuitive conclusion is that both paper and plastic cups can be preferable to ceramic cups, depending on consumption practices. This raises interesting systems questions: Are long-lasting multi-use products always environmentally preferable? Should complex products such as automobiles or computers always be manufactured for multiple reuse? What are the effects—on the economy, on technological choices, on company operations, on consumer options—of what seem to be environmentally beneficial changes in existing products or the introduction of "green" products and processes?

As with all systems work or analysis, definition of the systems boundaries is critical. This is the case both in biological ecology and in the analysis of engineering systems such as computer networks, transportation systems, or chemical processing facilities. The same challenge exists in trying to understand industrial ecosystems. For example, an industrial ecosystem may be defined by a single product. Any product, at a certain time, has a unique ecosystem characterized by raw material suppliers or component manufacturers, delivery, maintenance and collection systems, waste handlers, recyclers, and consumers. The various actors in industrial systems—raw material supplier or component manufacturer, consumer, waste handler, or recycler—are analogous to biological organisms. In these complex spatial and temporal webs of human production and consumption activities, individual materials may be traced through several different industrial ecosystems, and each industrial sector (and even company) may be characterized as playing a role in several industrial ecosystems. Alternatively, an industrial ecosystem may be bounded by geography (such as an urban area), an industry (such as agriculture), or a material (such as lead). Finally, just as from a global perspective, it is possible to think of the earth as made up of numerous interrelated natural ecosubsystems, so we may speak of the industrial ecosystem in terms of the whole network of industrial ecosubsystems.

Industrial ecology recognizes the unique role of humans in creating complex artifacts and institutions that force changes in materials and energy flows in both industrial and natural systems. Natural ecosystems, which provide the raw materials for economic activity, also serve as sinks for wastes from producers and consumers in industrial activities. The full, intricate, and complex set of interactions between industrial and natural ecosystems, however, is beyond the scope of this book, the subject of which is the value of using concepts derived from the study of natural ecological systems to understand interwoven natural and industrial systems. Thinking in this way—about industrial ecosystems—provides opportunities to examine and inform the ways in which producer and consumer practices in the economy may be altered to create environmentally compatible industrial ecosystems. Industrial ecology system boundaries may be drawn to include interacting industrial and natural processes. This book focuses primarily on the industrial components of these complex ecosystem extensions.

This volume of papers, drawn from those presented at a July 1992 National Academy of Engineering Workshop on Industrial Ecology and Design for Environment (DFE), represents an effort to advance the understanding of industrial ecology and to explore how companies can improve the environmental performance of their products, processes, and operations based on that understanding. In doing so, it builds on earlier efforts by the Academy, including *Technology and Environment* (Ausubel and Sladovich, 1989), *Energy: Production, Consumption, and Consequences* (Helm, 1990), and *Keeping Pace with Science and Engineering: Case Studies in Environmental Regulation* (Uman, 1993). Industrial ecology is still a young and evolving concept, and its ability to provide the theoretical or empirical underpinnings for action in companies, or by regulators or customers, is growing rapidly. However, what is known about the current ecology of industrial systems and their transformation suggests strategies and research efforts worth considering.

INDUSTRIAL ECOLOGY—A SYSTEMS APPROACH

Industrial ecology offers a unique systems approach within which environmental issues can be comprehensively addressed. It is based on an analogy of industrial systems to natural ecological systems.

> The idea of an industrial ecology is based upon a straightforward analogy with natural ecological systems. In nature an ecological system operates through a web of connections in which organisms live and consume each other and each other's waste. The system has evolved so that the characteristic of communities of living organisms seems to be that nothing that contains available energy or useful material will be lost. There will evolve some organism that will manage to make its living by dealing with any waste product that provides available energy or usable material. Ecologists talk of a food web: an interconnection of uses of both organisms and their wastes. In the industrial context we may think of this as being use of products and waste products. The system structure of a natural ecology and the structure of an industrial system, or an economic system, are extremely similar (Frosch, 1992, p. 800).

There are obviously limits to this analogy, but it can help illuminate useful directions in which the system might be changed. Consider, for instance, waste minimization at a scale larger than that of a single unit or facility in light of the biological analogue. A mature natural ecological community operates as a waste minimization system. In general, the waste produced by one organism, or by one part of the community, is not disposed of as waste by the total system so long as it is a source of useful material and energy. Some organism, some part of the ecological system, tends to evolve or adjust to make a living out of any particular waste. Microorganisms themselves are often turned into food for some other organism, and so the materials and the energy embedded in them (within the constraints of the second law of thermodynamics and given solar insolation) tend to circulate in a large, complex web of interrelated organisms.

In an industrial ecology, unit processes and industries are interacting systems rather than isolated components. This view provides the basis for thinking about ways to connect different waste-producing processes, plants, or industries into an operating web that minimizes the total amount of industrial material that goes to disposal sinks or is lost in intermediate processes. The focus changes from merely minimizing waste from a particular process or facility, commonly known as "pollution prevention," to minimizing waste produced by the larger system as a whole. It is not a new idea that waste should be considered a potentially useful material. A number of industries and industrial systems and processes are noted for using the waste of one process as feedstock for another. For example, steel scrap recycling is almost as old as the steel industry itself. The chemical and petrochemical industries characteristically attempt to convert as much of their raw material as possible into valuable products by finding uses for materials that were previously discarded, or making chemical production process changes so that waste materials can become products.

There are several barriers to finding new uses for waste, or changing processes so that waste generated has some value to a customer elsewhere in the industrial system. First, the data available to begin assessing the potential for recovery of useful by-products from waste are scanty and of poor quality, and need improvement. Current data do, however, illustrate the potential of prospecting in waste streams for highly concentrated, high-value material. In some instances, materials occur at higher concentration in waste streams than in natural ores. Moreover, because the potential value of recoverable material in the waste stream increases with concentration, it is often feasible to design waste streams for reuse and recycling (Allen and Behmanesh, in this volume).

Second, the incentive to find resources in waste streams depends in part on there being a reliable market for the waste by-product. Materials that have a high economic value are highly conserved. The automobile industry has long recycled components, such as starters and alternators, that would otherwise be waste. This use of potential waste occurs because there is a market for what would otherwise be disposed of as waste. The role of markets in promoting recycling is highlighted by the fact that 75 percent, by weight, of a car is currently recycled (Klimisch, in this volume). This is not the result of any mandate to recycle automobiles, but rather occurs because there is a market for the recycled material and for refurbished components and parts. The auto recycling infrastructure is one among many environmentally sound recycling systems that exist today. Care should be taken to ensure that public policy initiatives to encourage resource recovery do not throw such existing systems of materials recovery and reuse into disarray. New policy initiatives should focus on stimulating markets for material currently being disposed of as waste.

Past experience with recycling in the United States may provide insights into current recycling programs. In the 1930s and 1940s there was an economically viable paper recycling industry that was not motivated by environmental concerns. An investigation into its decline might prove instructive. The experience of the 1980s, when old newspapers collected for recycling were warehoused for lack of a market for the used paper, shows the futility of programs organized solely for the sake of recycling, independent of demand. An apparent, gradual market adjustment in the 1990s may eliminate the need for warehousing newsprint, but if this adjustment is inadequate, consideration will have to be given to other possible changes in paper recycling.

Third, there are clearly information deficiencies that hinder the operation of markets for waste and recycled material. Frequently lacking, for example, is information identifying who has what (supply), who needs what (market), who could use what (potential market), or who could produce something if somebody else wanted it (potential supply). These data are not available, because companies are frequently secretive about the composition of their waste streams (for fear that knowledge about waste streams can lead to deductions about proprietary processes), because different sectors are not familiar enough with each others' operations to know what opportunities are available, and because materials derived from waste streams are often considered inferior to virgin material.

Finally, the current regulatory structure can also prevent the linking of industries or industrial processes for more efficient use of waste materials (or recycled material). Some current regulations thwart rather than facilitate recovery and use of waste. As a result regulations often increase rather than decrease the amount of waste produced. Pfahl (in this volume) provides an example of how the ecologically beneficial recycling of lead dross would have been halted had a U.S. Environmental Protection Agency ruling labeling solder dross as hazardous waste been maintained. Recycling would have ceased because most of the secondary smelters involved in recycling lead from solder dross were not licensed to handle a material defined as hazardous waste under the Resource Conservation and Recovery Act, although it is not clear that any real hazard would have been posed by their handling of the dross. This type of problem arises not because anybody intended the outcome. Rather, the current media-specific environmental regulatory system focuses on the disposal and treatment of waste, not on minimizing waste. Thus, it can discourage attempts to reuse waste rather than provide incentives to do so.

Impediments to implementing environmentally preferable alternatives, however, are not restricted to environmental laws and regulations. For instance, antitrust laws, especially in the United States, can hinder industry cooperation that would be critical to developing comprehensive product and material recycling systems (Anderson, in this volume); inappropriate standards and specifications by large customers, such as the federal government, can also stifle the diffusion of environmentally preferable technologies (Morehouse, in this volume); and consumer protection laws that classify "remanufactured" products as "used" (secondhand) discourage product-life extension activities such as refurbishment and component reuse.

A broader, systems-based approach to industrial activity thus begins to illustrate the deficiencies in current understanding of industrial ecologies and the consequences and effects of changing production and consumption processes. It also exposes the impact of fiscal and regulatory policies in shaping the structure and operation of industrial ecosystems. There is clearly a need to better understand current industrial ecosystems and to tap points of leverage within them to improve the environmental performance of these systems.

Several papers in this volume refer to sustainable development. Sustainable development represents the quest for an economy that exists in equilibrium with the earth's resources and its natural ecosystems. Sustainable development brings environmental quality and economic growth into harmony, not conflict. It is a concept that recognizes that economic activities and environmental considerations need to be integrated for humanity's long-term well-being. Sustainable development, however, is framed in broad and vague terms of restructuring social, economic, technological, and industrial policies and practices, of building new institutions, and of controlling population. Industrial ecology hints at concrete steps that lead to continuous environmental improvement that is also economically beneficial.

The current ecology of industrial systems suggests several environmentally desirable changes in industrial production and practices. These changes include improving the efficiency and productivity of industrial systems, minimizing waste in the use of raw materials, substituting abundant and environmentally benign materials for those that are less so, developing uses for waste products, and reusing manufactured products at the end of their first life. The goal of efficient materials and energy use suggests exploring ways in which the web of waste recycling and reuse found in natural systems may be imitated in the industrial context. Moving toward such a system, as illustrated earlier, involves changing a complex mix of interacting factors. To do so it may be necessary to improve current information systems or create new ones, remove regulatory barriers, and devise economic incentives.

INDUSTRIAL ECOLOGY AND ITS CONTEXT

Understanding Industrial Ecology

To chart a course toward environmentally preferable industrial systems and practices, it is necessary to consider the current state of industrial ecosystems and to set goals for the use of materials and energy to achieve the ideal state. In addition it is necessary to explore policies and incentives that will reorient production and consumption systems through natural market forces.

It is useful to consider three possible stages in the evolution of an industrial ecosystem. Type I industrial ecosystems (Figure 1), are characterized by linear, one-way flows of materials and energy where the production, use, and disposal of products occur without reuse, or recovery, of energy or materials. In Type II industrial ecosytems (Figure 2), some internal cycling of materials occurs, but there is still a need for virgin material input, and wastes continue to be generated and disposed of outside the economic system (i.e., as emissions to air or water). Hypothetical Type III industrial ecosystems (Figure 3) would be characterized by complete or nearly complete internal cycling of materials. Material is highly conserved, no waste material is released, and heat escapes. Type III industrial ecosystems are most similar to natural systems. The energy and materials flows provide some fundamental insights.

Like the earth's biota, all three industrial ecosystems models depend on continuous inputs of energy. This raises two questions: What are the prospects for meeting energy needs in the long term? and Can the environmental impacts of energy conversion be minimized? Linden (in this volume) suggests that fossil

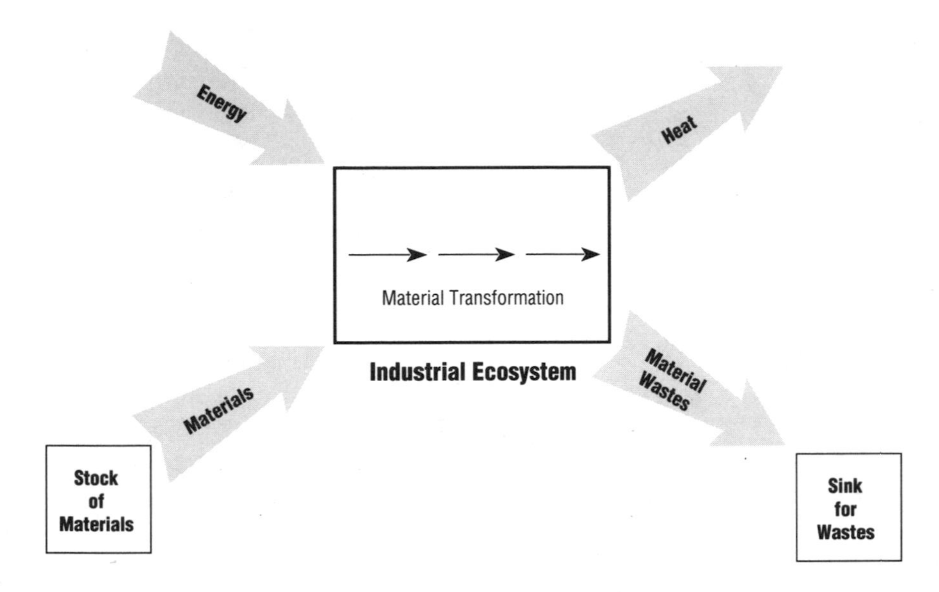

FIGURE 1 Type I industrial ecosystem.

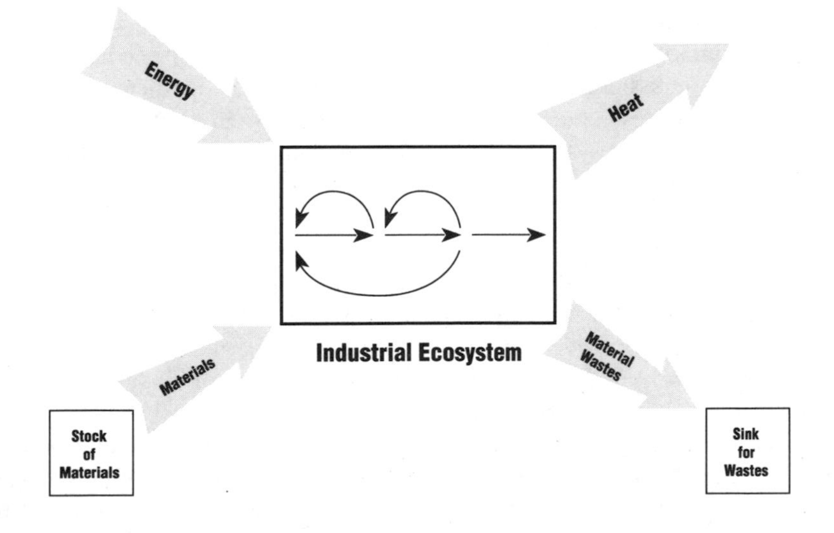

FIGURE 2 Type II industrial ecosystem.

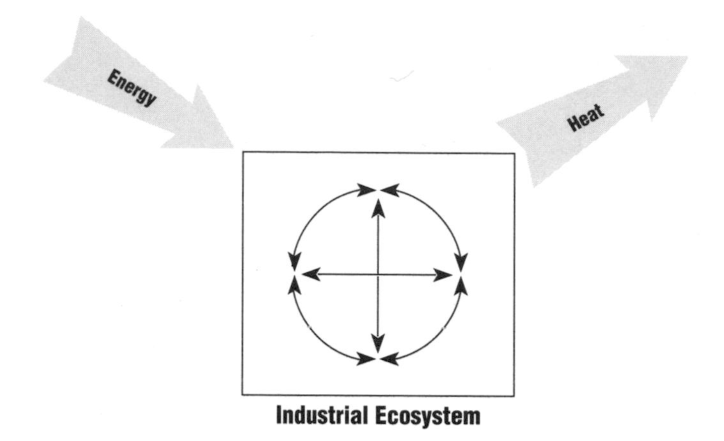

FIGURE 3 Type III industrial ecosystem.

energy reserves are adequate to support a rational and planned transition to sustainability. However, emissions from the use of such resources—particularly heavy metals and carbon dioxide—will have to be controlled if this potential is to be exploited in environmentally acceptable ways.

Probing deeper we find that this is but part of the answer. Duchin (in this volume) suggests that the geographic locus of emissions of carbon dioxide and oxides of sulfur and nitrogen will continue shifting from the rich to the poor economies, even if only moderate economic growth is achieved in developing countries over the next several decades. At the same time, there will be a significant increase in the total emissions of carbon dioxide. These shifts are likely to occur even under optimistic scenarios of accelerated adoption of modern, commercially proven technologies to reduce and control pollution in both rich and poor nations. Therefore, if emissions are to be held at current levels, significant changes in production and consumption practices will be required. In addition, strategies to improve industrial ecosystems in industrialized countries are likely to differ from those in developing nations.

Current industrial ecosystems can be characterized by a mix of Type I and Type II materials flows. Industrial cycles tend to be open, with little recycling. The Type III industrial ecology model, which most closely resembles natural ecosystems, is a "no waste" ecology. It is in keeping with the limiting goal of "zero discharge" adopted by several major companies. Yet, the elimination of manufacturing wastes, or zero discharge, is beyond the capability of modern technology and would require the full participation of consumers. It is probably an unattinable goal, but in pursuing that objective it is important to recognize that there are only two possible long-run fates for materials—dissipative loss and recycling or reuse. Recycle and reuse will therefore have to be a part of any quest for a zero-discharge industrial ecology.

Ayres (in this volume) suggests two metrics that can be used to measure the extent to which materials cycles have been closed. The first, a surrogate measure of "distance" from a steady-state condition, is the fraction of current metal supply needed to replace dissipative losses (i.e., production from virgin ores needed to maintain a stable level of consumption). The second is a measure of materials productivity, which is the economic output per unit of material. Materials productivity can be measured at various scales—for the economy as a whole, for each sector, or for each firm—as well as for each major element, such as carbon, oxygen, hydrogen, sulfur, chlorine, iron, or phosphorus. These types of metrics can be useful in gauging the environmental efficiency of production and consumption practices based on the flow of materials.

The study of the flow of materials within and among industrial systems can also reveal opportunities for material recovery and reuse. Allen and Behmanesh (in this volume) give a simplified model of lead circulation through the economy. Using data on production, consumption, and waste flows, they find (not surprisingly) that waste streams with high concentrations of lead are more likely to be recycled than dilute waste streams. One would also expect to find other raw materials of high value in waste streams. Yet, even though the concentrations of metal resources in many waste streams are higher than for typical virgin resources, much of this waste material is now being discarded. Reductions in the quantity of material being jettisoned may be achieved through improvements in information about these hidden opportunities, more extensive trading in waste, appropriate regulatory reform, and advancements in separation and recovery technologies.

The Context of Industrial Ecology

The flow of materials, the use of energy, and the development and application of technology to produce and use goods and services do not occur in a vacuum. Each of these interrelated aspects of economic activity is influenced by a complex mix of dynamic economic forces, business practices, technological options, regulations, scientific understanding, and public opinion. Social values, and public perceptions and demands, often shape strategies to address public interest concerns such as the environment. Regulations, for example, reflect public values and are crafted to address societal concerns. Their structure, however, can be more or less efficient in achieving the desired social end. Currently, the majority of environmental regulations are rigid and prescribe methods of controlling emissions rather than encouraging desired outcomes. Housman (in this volume) argues that command-and-control regulations may have reached the limits of their effectiveness and must be replaced by more market-oriented strategies that provide the same or a higher level of environmental protection. Such a movement, however, will require that market-oriented regulations still protect the values society places on the possible outcomes or results.

The value society places on natural resources, however, is difficult to gauge and is not static. It changes as a result of economic and cultural factors. For example, in both rich and poor nations, expenditures to achieve environmental gains must be justified against the need for resources to fight poverty, disease, and ignorance. Richer nations are likely, however, to have more resources to spend on environmental protection than poorer countries. Societal values can also be altered by the availability of management and technological solutions. Social costs associated with the environment, therefore, can be influenced by the substitutability of technology and knowledge for natural resources.

Crosson and Toman (in this volume) present four limiting scenarios relating substitutability and social costs: (1) complete substitutability among resources and zero social costs; (2) complete lack of substitutability and zero social costs; (3) complete lack of substitutability and catastrophic social costs; and (4) complete substitutability and catastrophic social costs. Markets may be effective for selecting among available options where there are many substitutes that can be exploited with low social cost trade-offs among resources. However, where there are few substitutes for resources and the social cost of exploitation is high, direct government intervention may be needed. Decisions to limit free market forces to address environmental solutions occur when a particular situation is judged to be too risky to leave to market forces. Such judgments, however, may vary widely among the individuals and groups involved, and the challenge is in employing a mix of regulations and incentive mechanisms to achieve desired outcomes. The newer initiatives such as pollution prevention and the precautionary principle (Housman, in this volume) seek to avoid environmental harms before they occur. By setting targets instead of dictating solutions, they allow the regulated community to compete to meet the goals. They are examples of ways to encourage internalization of environmental costs into the market. Environmental quality is often considered to be free or underpriced by the marketplace. The aim of these initiatives is to have the market prices for goods and services reflect a proper valuation of environmental impacts associated with their production, use, and disposal.

The internalization of environmental costs through public policy decisions currently occurs through direct regulation or (to a lesser extent) through economic instruments. Each approach has advantages and disadvantages. Properly designed regulations can produce swift and relatively predictable results (as was the case with the mandatory phaseout of leaded gasoline), but they can also impose unnecessary costs on industry and stifle environmentally innovative designs; for example, they can discriminate against new technologies by prescribing specific technologies as rigid standards (such as so-called best available emissions-control technology) (Weinberg et al., in this volume). In addition, prescriptive environmental regulations are unfortunately too often written as if the world were static,[1] and do not consider the risks to society of not innovating.

Economic instruments, including surcharges on industrial emissions and taxes on undesirable compounds such as ozone-depleting substances, can provide flexibility by providing direction toward desired outcomes rather than dictating methods to achieve them. They are suited to addressing dissipative environmental problems, such as groundwater contamination or heavy metal dispersion into the atmosphere. These environmental problems are not linked to large point sources of pollution and therefore are not amenable to being contained by command-and-control methods. Economic instruments are also better suited to influencing product design for environmental benefits. The proliferation of design and materials technology choices, the global nature of production and consumption (with materials, parts, and components frequently crossing many borders in both the assembly and the sale of a product), and the multidimensional nature of product impacts require flexible policies. While economic instruments offer several advantages, they can be expensive to administer and are often politically unpopular. The challenge for policymakers is to employ a mixture of regulations and economic instruments that encourage designers to take account of rapid technological change while simultaneously safeguarding environmental quality (Weinberg et al., in this volume).

EMERGING COMPANY PRACTICES

Even as policymakers grapple with ways to encourage environmentally sound production and consumption practices, private firms are changing practices in response to competitive and environmental pressures. World-class manufacturing practices that have emerged in recent years call for a systems approach to managing the complex manufacturing enterprise.[2] Because the environment is a critical operating component in manufacturing, environmental issues are beginning to be considered in the product realization processes and operations of companies together with traditional criteria such as cost, quality, and performance.

Two manufacturing practices that facilitate the integration of environmental factors in industrial production are concurrent engineering and total quality management (TQM). First, environmental factors can be considered as an element of the concurrent engineering "Design for X" (DFX) process, where "X" represents a product characteristic such as reliability or manufacturability that the company wants to maximize in its product design (Allenby, in this volume). Second, the adoption of TQM programs has made quality the responsibility of everyone in the company, not just the organization's internal quality group. By using existing or developing TQM programs, it is possible to translate environmental responsibilities across the board in a similar manner (Pfahl, in this volume). Both concurrent engineering practices and TQM techniques seek to integrate as many external factors as possible into initial design activities and thereby avoid problems that may occur after the fact. As environmental considerations are cast as critical components of concurrent engineering practices or as important quality issues, they are more easily accepted as standard practice within companies.

Several features common to world-class manufacturing practices and to integrating environmental factors in companywide operations are worth noting.

- Senior management commitment to change
  In the larger firms prominent in advancing environmentally preferable practices (or at least in publicizing them), there typically is a commitment of senior management to environmental progress (Dillon, in this volume). Such commitment is necessary to achieve fundamental changes in culture, organization, and procedures.
- Goals, priorities and strategies
  Corporate goals, priorities, and clear strategies to achieve goals are the basis of good project planning and, not surprisingly, are critical in efforts to effect change in corporations.
- Use of cross-functional teams
  Both concurrent engineering and TQM make use of cross-functional teams to ensure sensitivity to changes in the external climate (customer needs, and regulatory, technical, legal, social, as well as environmental requirements). Product development teams include the traditional product managers, design engineers, and individuals with expertise in manufacturing, marketing, sales, law, and the environment.
- Learning from current product lines
  An examination of current product lines and production practices can reveal opportunities for environmental improvement. Baselining can be a useful exercise to determine how well a product meets certain environmental criteria and to evaluate the environmental effects associated with the raw materials chosen by the design team. It can also reveal ways to reduce those effects (Sekutowski, in this volume). Cross-functional teams can be effective in evaluating products, processes, and operations and identifying opportunities for continuous improvement, including environmental improvement.
- Partnering to achieve common goals
  There are several ways in which companies and other organizations can work together to meet common goals, including environmental goals. Partnerships may be formed horizontally between companies within the same industry to achieve common noncompetitive aims or vertically with suppliers, vendors, and recyclers to ensure that environmental impacts of a product's production, delivery, sale, use, maintenance, recycle, reuse and disposal are handled appropriately (Klimisch, in this volume). Alliances can also be formed with universities and government laboratories to support related studies and to tap expertise not necessarily available within companies. Productive alliances can also be forged with environmental advocacy groups.
- Managing the supplier chain
  The information found in supplier management systems is an indirect systems component that can affect design decisions in a firm. Suppliers can be

asked to meet specifications that are environmentally appropriate while still maintaining necessary quality, cost, and performance standards. For instance, military specifications and standards have an extensive influence on global manufacturing practices and component selection practices (Morehouse, in this volume). A change in these specifications, therefore, can have a profound effect on practices of companies that supply the military.

A private firm's standards and specifications can have a similar effect on the design of its products and operations, as well as on the environmental performance of its suppliers. Private firms can, therefore, ensure that their products are environmentally superior by reviewing their standard components lists, which identify stock components to be used in design where possible, and ensuring that all recommended components are as environmentally acceptable as possible. These firms can also use their purchasing contracts to influence suppliers' behavior by requiring environmentally preferable practices.

In addition to these practices, the environmental life cycle approach (Allenby, in this volume), total environmental cost accounting (Todd, in this volume), and the selling function of the product (Stahel, in this volume) raise issues regarding the reorientation occurring in firms that respond to environmental concerns.

Taking a Life Cycle Approach

The most far-reaching implication of integrating environmental concerns in the economic decisions of companies and society is the need to take a life cycle approach to environmental analyses. This approach requires that environmental impacts—with "environmental" taken broadly to include relevant safety, health, and social factors—be understood and summed up across the lifetime of the product, process, material, technology, or service being evaluated. The goal is to reduce to a minimum the overall environmental impact of a product or process, and not simply address one aspect of that impact. This goal becomes important because minimizing the impacts of subsystems does not ensure that the impacts of the entire system are minimized, or even reduced.

Most judgments of environmental preferability hinge on life cycle assessments (LCA) of products, material, and processes and on the actions taken.[3] These assessments provide a comprehensive profile of energy and material inputs to, and environmental impacts of, making, using, and retiring products. However, LCAs are seldom able to provide a clear-cut answer on environmental preferability or to provide a definitive measure of product "greenness." Studies of fairly simple consumer items, such as reusable vs. disposable diapers, have come to opposite conclusions depending on the assumptions made. The assumptions have to do with the types and modes of pollution and energy use considered in the analysis. If it is difficult to make an environmental distinction between different types of diapers, cups, or bags (products of simple design, minimal material diversity, but high material density), the challenge the United States Air Force faces in defining "green weapons," (which are of complex design, contain a complex mix of materials, and have a comparatively low material density) is clearly beyond the current state of the art.

The notion of "green" weaponry (discussed by Morehouse, in this volume) is often greeted initially with bemusement. However, it raises some challenging questions: what are the environmental implications of maintaining aircraft, ships, and other weapons platforms over many years (frequently beyond design lifetimes), and how can the environmental impacts be reduced through more appropriate design? More fundamentally, if all aspects of a life cycle are to be considered, and consistent answers are not possible in the comparison of "simple" articles such as cups, what does it mean to designate a more complex product—an aircraft, a ship, an automobile, a computer—as environmentally superior?

If the environmental preferability of products cannot be determined unambiguously—and it is doubtful that the methodologies and sufficient data to identify environmentally preferable options do generally exist—regulators, design teams, and product and process engineers face a serious conundrum. They are being asked to promote and develop environmentally desirable materials, technologies, processes, and products with little valid guidance on what "environmentally preferable" means in practice, how these choices can be identified, or how their choices may affect other parts of industrial ecosystems, including raw material suppliers or component manufacturers, delivery, maintenance and collection systems, waste handlers, recyclers, and consumers.

The comprehensive life cycle assessment is data intensive and can vary depending on the quality of data available, the biases of the assessor, and the assumptions made. The need, however, is for simple methodologies that are inexpensive to use. As a corollary, no one has yet developed a methodology that easily identifies first-order environmental effects and separates them from the innumerable second- and third-order effects that any design choice entails. There is, therefore, much work to be done in refining life cycle methodologies and in bridging the gaps in information. In the final analysis, it is a question of choosing from among several alternatives—one or another design emerges from among different designs. If a selection is to be made among several alternatives that equally meet the traditional criteria of cost, quality, function, and customer acceptance, better environmental life cycle information may tip the balance in favor of an environmentally preferable design.

Most LCA methodologies have been developed in the context of chemical substance risk assessment, and some of the proposed methodologies appear better suited to relatively simple products, such as consumer personal care products or plastic packaging, rather than complex manufactured durable goods. The Design for Environment (DFE) methodologies, proposed by Allenby (in this volume) illustrate practical applications of life cycle approaches to environmental decision making in complex manufacturing operations, and ways to apply them to subassemblies or completed products. The DFE methodologies can be developed as a module of concurrent engineering practices or the product realization process that can facilitate the integration of environmental objectives and constraints in the design process directly. At the same time, there is a need to improve tangential systems, such as the accounting or supplier management systems, which affect design decisions indirectly.

Getting at the "Total Environmental Costs" in a Firm

Existing accounting systems can prevent modern firms from internalizing environmental costs and considerations and can compound difficulties encountered in effecting environmentally preferable changes. Environmental costs that are buried in overhead accounts are hidden from managers. As a result, these costs are not seen and cannot be controlled. The need to improve accounting systems is no trivial matter. Incentives (and disincentives) must be fed back to the appropriate decision makers.

Accounting systems must capture currently hidden and unaccountable costs, such as product-related legal expenses, regulatory costs, public relations expenses, and the opportunity costs of failing to adopt clean technologies (Todd, in this volume). There are, however, severe institutional barriers that prevent managers from getting the information necessary to pursue optimal environmental waste-reduction strategies. First, current accounting systems were not designed to capture much of the engineering and accounting data required for environmental decision making. Second, data that are collected and processed are usually aggregated in such a way that they lose their environmental information content (as well as managerial control). Finally, line managers are rarely made responsible for environmental costs.

If accounting practices are to be changed, important institutional barriers must be removed, and incentives must be provided to motivate aggressive and creative development of solutions to reduce or eliminate wasteful processes and practices. These steps are important because the compensation of managers is frequently based in part on profitability that results from reducing controllable costs. Managers, therefore, have strong disincentives to seek "full environmental costing" within the firm, because this would bring additional costs under their accountability. The paradox is that managers cannot act without adequate information, and yet they will not act voluntarily if the result is not recognized and may in fact hurt the measure of their performance.

Shifting to Selling the Function of the Product

From a sustainable development perspective, industrial ecology suggests that the flow rate of materials in the economy may have to be slowed. Stahel (in this volume) advocates a shift to an economy that sells the functionality of products rather than the products themselves to decelerate materials flow. This concept implies, for example, offering pesticide control instead of pesticides, communication services instead of phones, computing power instead of computers, refrigeration instead of refrigerators, and transportation instead of automobiles. It suggests valuing products not by the cost of their production but by the value the customer derives from using the products. An economy based on providing "functionality" would encourage reuse of products over recycling of products.

This idea in fact reflects both a past practice and a growing trend. For instance, it was not long ago that the Bell Telephone System in the United States owned the phones its customers used and then recycled or reused the phones as customers turned them back in. There is also evidence of some companies experimenting with providing functionality of products in their marketing efforts. For example, a new business group at Dow Chemical, Advanced Cleaning Systems, intends to maintain and increase the sales and profitability of the company's chlorinated solvents business, which is threatened by increased regulation of ozone-depleting chemicals and federal and state programs to control toxic air emissions (Dillon, in this volume). The group is developing not only alternative chemicals and processes but a range of customer services as well—improved process controls and the recycling of spent chemical solvents. It offers delivery of new chemicals to its customers in conjunction with "take-back" of the used chemical in reusable containers (Dillon, in this volume). The spent chemical can then be cleaned or reprocessed and be either returned to service or disposed of appropriately. This practice is an example of asserting control over the product life cycle while providing the customer with the functions of the product.

In the longer term, the possibility of a "functionality economy" raises truly revolutionary implications, not just for manufacturing firms and their customers but for society and regulators as well. One obvious implication is that products would be built to last long. Yet, because older, less efficient products such as automobiles may pollute more, manufacturers may have to design them to be easily upgradable. The more difficult issues of a functionality economy involve figuring out how to handle the impact of extended manufacturer liability for products and how to evaluate the costs of providing the service as well as the associated value of the service provided.

EDUCATION NEEDS AND RESEARCH DIRECTIONS

Universities have a unique role to play in influencing the evolution of industrial ecosystems. In education, there is a need to base concepts such as pollution prevention, waste minimization and reuse, as well as product and material recycling and reuse considerations on broad engineering science principles, and to introduce them into engineering education. This can be accomplished by showing how environmental considerations of key design features can be balanced with other design factors such as cost, aesthetics, functional performance, reliability, and quality. Friedlander (in this volume) suggests that this is one of the greatest challenges currently facing the engineering profession.

New engineering curricula may have to be developed to educate the next generation of engineers so that they can incorporate environmental factors into design decisions. New courses may be needed, but it may be more effective to integrate some of the lessons learned from industry case studies (and joint university-industry research) into existing design classes across the various engineering disciplines. In addition, engineers also need a good understanding of the interaction between the technological and environmental systems at various scales—macro (level of the economy), meso (level of the firm or plant), and micro (the level of the process)—of industrial organization. Finally, environmental considerations must be integrated into the curricula of other disciplines, such as business, law, economics, and public policy. Changes needed in education can be accelerated if industry clearly articulates the qualifications it seeks in the graduates it hires.

In research and policy, academic research institutions can tackle many of the pressing questions about industrial ecology and ways to create sustainable industrial ecologies. Examples of research needs include work on engineering and economic systems analyses and forecasts of industrial metabolism, input-output models of the economic and environmental implications of alternative strategies, development and analysis of waste stream composition data to target waste recovery and reuse, technologies to recover material from waste streams, tools to aid environmentally based material selection in design, management strategies for life cycle product management, and management studies of corporate cultural change. Universities are also in a unique position to play the role of broker in contentious policy debates and to convene groups with diverse views and opinions to shape policy on neutral grounds (Ehrenfeld, in this volume).

SUMMARY

The flow and embodiment of materials and energy in production and consumption activities of the economy underlie both economic growth and environmental perturbations. The familiar remedy for environmental ills has been to address the symptoms of pollution—in the air or water and on land—at the end of production and consumption processes. Command and control at the end of such processes has had little effect in altering their internal workings or the composition of products and did not consider the operation of industrial ecosystems or the influence of economic, social, and political pressures. As the shift is made from controlling pollution to preventing it and, beyond that, to achieving sustainable development, comprehensive approaches that focus on economic systems and the flow of materials and energy are needed to address the complex mix of issues raised about energy use, material choices, product and process design, interfirm relations, material and waste management, market responses, information needs, and public policy choices.

Industrial ecology provides a systems perspective based on its analogy to biological ecosystems. This analogy illustrates the points of leverage for, and barriers to, improving the environmental characteristics of industrial ecosystems at the macro-level (the global economy), meso-level (the firm), and the micro-level (the industrial process). At the macro-scale the barriers include the inadequacy of current data about wastes and their potential recovery, the absence of reliable markets for waste by-products, the piecemeal environmental and fiscal regulatory approaches that inadvertently discourage environmentally sound practices, divergent values held by stakeholders, population growth, and significant income disparities among countries. A variety of structural and stakeholder factors affect industrial ecology at the macro-levels, but industrial ecology offers a framework within which all may be evaluated.

Taken together, the authors in this volume offer substantial evidence that change is beginning to occur in the understanding of industrial ecology and that changes in practice are occurring in industry, the military, academia, and government to reduce the environmental impacts of economic development. Nonetheless, there are barriers to change in the structures of both government and private firms such that change will be neither quick nor easy. At the theoretical and applied level, it is apparent that a great deal of research is necessary to inform the decisions being made by governments, the actions being taken by companies, and the education of engineers, lawyers, business managers, sociologists, and economists. This book provides a basis for intensifying that effort.

NOTES

1. *Keeping Pace with Science and Engineering: Case Studies in Environmental Regulation* (edited by M. Uman), the National Academy Press, explores the challenges to environmental regulation posed by changes in scientific and technical understanding.
2. For more on principles, or foundations, that have proved effective in improving manufacturing systems, see *Manufacturing Systems: Foundations of World-Class Practice*, edited by J. Heim and D. Compton, National Academy Press, 1992.
3. A good example of such a methodology is given by the Society for Environmental Toxicology and Chemistry (1991). The Centre for Environmental Science (1992) provides a good overview of the status of life cycle assessment.

REFERENCES

Ausubel, J. H., and H. E. Sladovich, eds. 1989. Technology and Environment. Washington, D.C.: National Academy Press.

Centre for Environmental Science. 1992. Methodology for Environmental Lifecycle Analysis: International Developments. Paper prepared as a part of the National Research Programme for Recycling of Waste Products, J. A. Assies, investigator. University of Leiden, Netherlands.

Frosch, R. A. 1992. Industrial ecology: A philosophical introduction. Proceedings of the National Academy of Sciences 89(February):800-803.

Heim, J. A., and W. D. Compton, eds. 1992. Manufacturing Systems: Foundations of World-Class Practice. Washington, D.C.: National Academy Press.

Helm, J. L., ed. 1990. Energy: Production, Consumption, and Consequences. Washington, D.C.: National Academy Press.

Society for Environmental Toxicology and Chemistry (SETAC). 1991. A Technical Framework for Life-Cycle Assessments. Washington, D.C.: The SETAC Foundation.

Uman, M. F., ed. 1993. Keeping Pace with Science and Engineering: Case Studies in Environmental Regulation. Washington, D.C.: National Academy Press.

van Eijk, J., J. W. Neiuwenhuis, C. W. Post, and J. H. de Zeeuw. 1992. Reusable Versus Disposable: A Comparison of the Environmental Impact of Polystyrene, Paper/Cardboard and Porcelain Crockery. Ministry of Housing, Physical Planning and Environment, Deventer, Netherlands.

Understanding Industrial Ecology and Its Context

The Greening of Industrial Ecosystems. 1994.
Pp. 23–37. Washington, DC:
National Academy Press.

Industrial Metabolism: Theory and Policy

ROBERT U. AYRES

The word *metabolism*, as used in its original biological context, connotes the internal processes of a living organism. The organism ingests energy-rich, low-entropy materials (food), to provide for its own maintenance and functions, as well as a surplus to permit growth or reproduction. The process also necessarily involves excretion or exhalation of waste outputs, consisting of degraded, high-entropy materials. There is a compelling analogy between biological organisms and industrial activities—indeed, the whole economic system—not only because both are materials-processing systems driven by a flow of free energy (Georgescu-Roegen, 1971), but because both are examples of self-organizing "dissipative systems" in a stable state, far from thermodynamic equilibrium (Ayres, 1988).

At the most abstract level of description, then, the metabolism of industry is the whole integrated collection of physical processes that convert raw materials and energy, plus labor, into finished products and wastes in a (more or less) steady-state condition (Figure 1). The production (supply) side, by itself, is not self-regulating. The stabilizing controls of the system are provided by its human component. This human role has two aspects: (1) direct, as labor input, and (2) indirect, as consumer of output (i.e., determinant of final demand). The system is stabilized, at least in its decentralized competitive market form, by balancing sup-

NOTE: This paper is reprinted with minor editorial changes from *Industrial Metabolism*, R. Ayres and U. Simonis, eds., United Nations University Press, Tokyo, Japan, 1993,

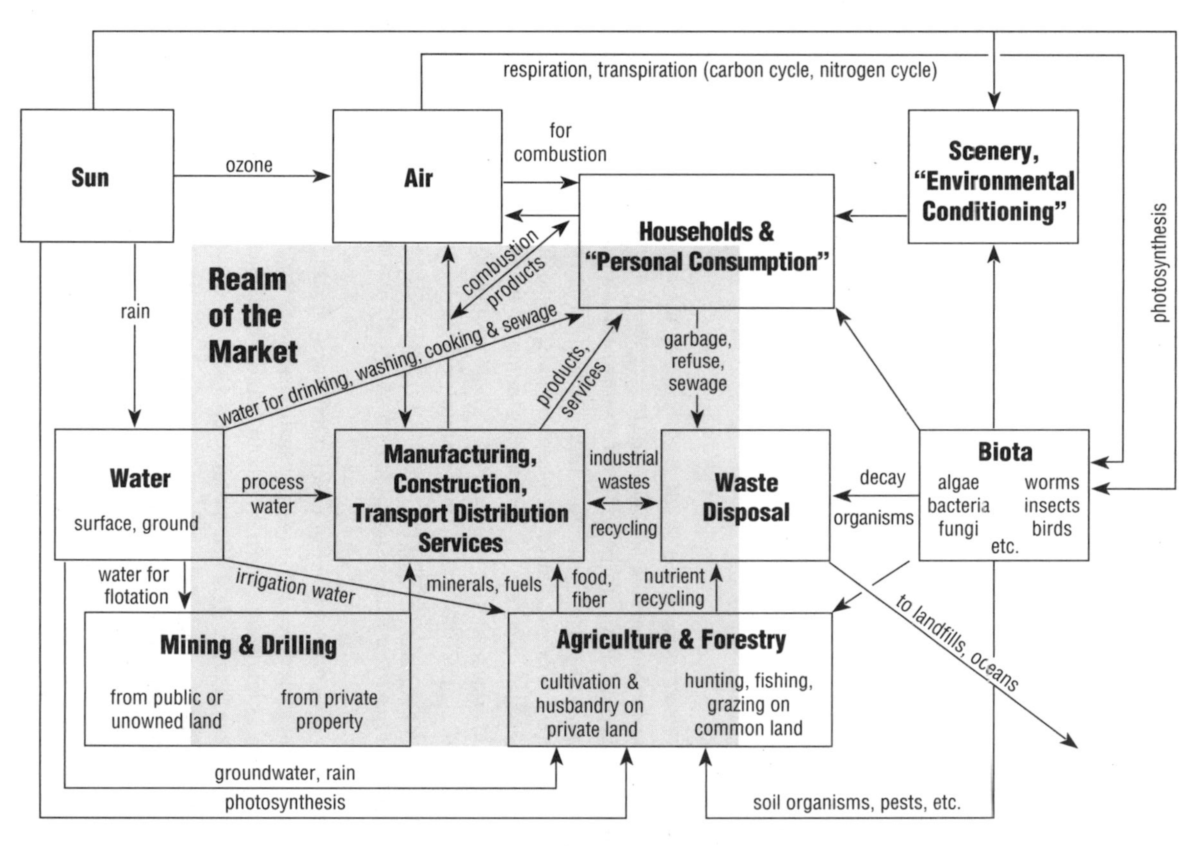

FIGURE 1 What is industrial metabolism?

ply of and demand for both products and labor through the price mechanism. Thus, the economic system is, in essence, the metabolic regulatory mechanism.

Industrial metabolism can be identified and described at a number of levels below the broadest and most encompassing global one. Thus, the concept is obviously applicable to nations or regions, especially "natural" regions such as watersheds or islands. The key to regional analysis is the existence of a well-defined geographical border or boundary across which physical flows of materials and energy can be monitored.

The concept of industrial metabolism is equally applicable to another kind of self-organizing entity, a manufacturing enterprise or firm. A firm is the economic analogue of a living organism in biology.[1] Some of the differences are interesting, however. In the first place, biological organisms reproduce themselves; firms produce products or services, not other firms (except by accident). In the second place, firms need not be specialized and can change from one product or business to another. By contrast, organisms are highly specialized and cannot change their behavior except over a long (evolutionary) time period. In fact, the firm (rather than the individual) is generally regarded as the standard unit of analysis in economics. The economic system as a whole is essentially a collection of firms, together with regulatory institutions and worker-consumers, using a common currency and governed by a common political structure. A manufacturing firm converts material inputs, including fuels or electric energy, into marketable products and waste materials. It keeps financial accounts for all its external transactions; it is also relatively easy to track physical stocks and flows across the "boundary" of the firm and even between its divisions.

THE MATERIALS CYCLE

A third way in which the analogy between biological metabolism and industrial metabolism is useful is to focus attention on the "life cycle" of individual "nutrients." The hydrological cycle, the carbon cycle, and the nitrogen cycle are familiar concepts to earth scientists. The major way in which the industrial metabolic system differs from the natural metabolism of the earth is that the natural cycles (of water, carbon/oxygen, nitrogen, sulfur, etc.) are *closed*, whereas the industrial cycles are *open*. In other words, the industrial system does *not* generally recycle its nutrients. Rather, the industrial system starts with high-quality materials (fossil fuels, ores) extracted from the earth, and returns them to nature in degraded form.

This point particularly deserves clarification. The materials cycle, in general, can be visualized in terms of a system of compartments containing *stocks* of one or more nutrients, linked by certain *flows*. For instance, in the case of the hydrological cycle, the glaciers, the oceans, the freshwater lakes, and the groundwater are stocks, while rainfall and rivers are flows. A system is *closed* if there are no

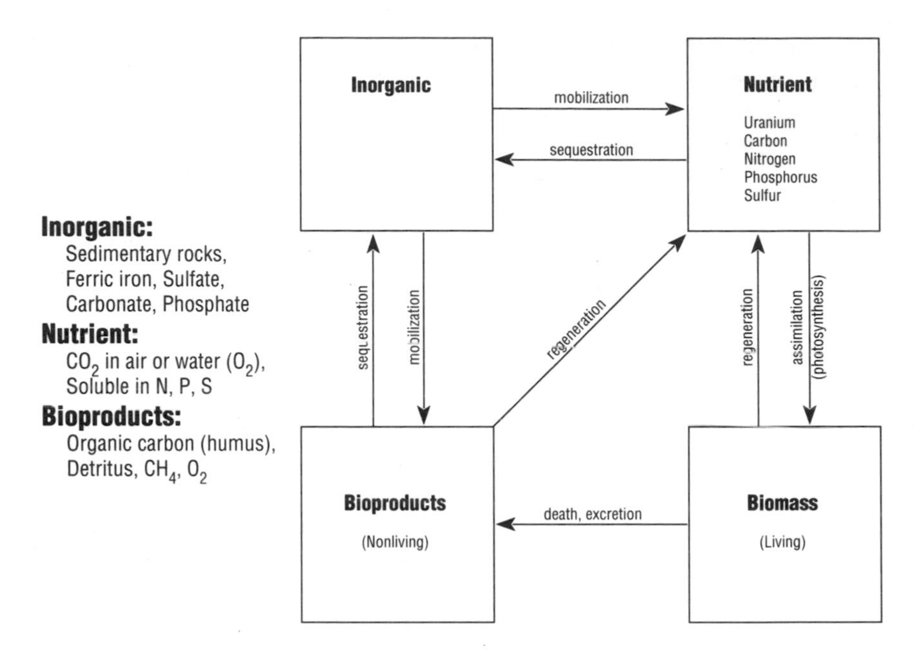

FIGURE 2 Box scheme for bio-geo-chemical cycles.

external sources or sinks. In this sense, the Earth as a whole is essentially a closed system, except for the occasional meteorite.

A closed system becomes a *closed cycle* if the system is also in steady-state, that is if the stocks in each compartment are constant and unchanging, at least on the average. The materials balance condition implies that the material inputs to each compartment must be exactly balanced (on the average) by the outputs. If this condition is not met, for a given compartment, then the stock in one or more compartments must be increasing, while the stocks in one or more other compartments must be decreasing.[2]

It is easy to see that a closed cycle of flows, in the above sense, can be sustained indefinitely only by a continuous flow of free energy. This follows immediately from the second law of thermodynamics, which states that global entropy increases in every irreversible process. Thus, a closed cycle of flows can be sustained as long as its external energy supply lasts. An open system, on the contrary, is inherently unstable and unsustainable. It must either stabilize or collapse to a thermal equilibrium state in which all flows, that is, all physical and biological processes, cease.

It is sometimes convenient to define a generalized four-box model to describe materials flows. The biological version is shown in Figure 2, while the analogous industrial version is shown in Figure 3. To revert to the point made at the beginning of this section, the natural system is characterized by closed cycles, at least

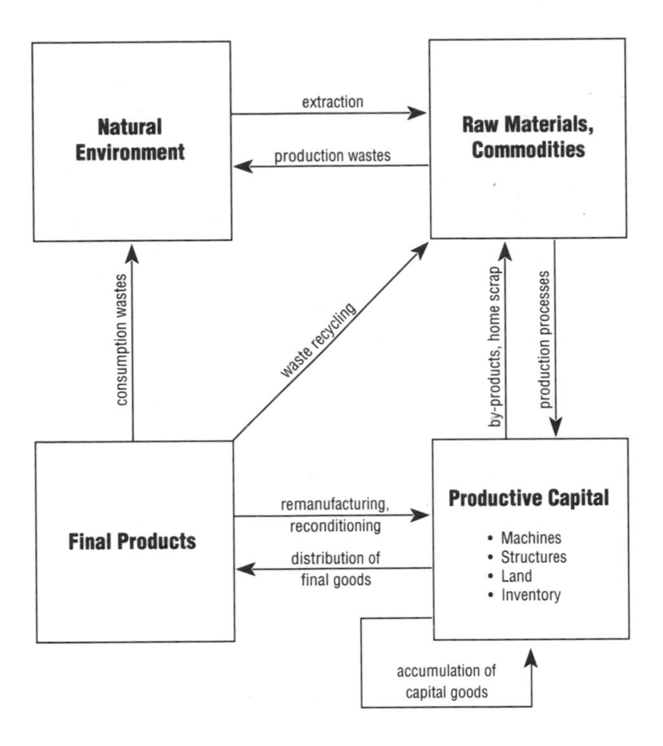

FIGURE 3 Box scheme for industrial material cycles.

for the major nutrients (carbon, oxygen, nitrogen, sulfur)—in which biological processes play a major role in closing the cycle. By contrast, the industrial system is an open one in which "nutrients" are transformed into "wastes" but are not significantly recycled. The industrial system of today is therefore unsustainable.

At this stage, it should be noted that nothing can be said about open cycles (on the basis of such simple thermodynamic arguments, at least) with respect to any of the really critical questions. These are as follows: (1) Will the industrial system stabilize itself without external interference? (2) If so, how soon, and in what configuration? (3) If not, does there exist any stable state (i.e., a system of closed materials cycles) short of ultimate thermodynamic equilibrium that could be reached with the help of a feasible technological "fix"? (4) If so, what is the nature of the fix, and how costly will it be? (5) If not, how long do we have until the irreversible collapse of the bio-geosphere system makes Earth uninhabitable? (If the time scale is a billion years, we need not be too concerned. If it is a hundred years, civilization, and even the human race, could already be in deep trouble.) It is fairly important to try to find answers to these questions. Needless to say, we do not aspire to answer all these questions in the present paper.

It should also be pointed out that the bio-geosphere was not always a stable system of closed cycles. Far from it. The earliest living cells on Earth obtained their nutrients, by fermentation, from nonliving organic molecules whose origin is still not completely understood. At that time the atmosphere contained no free oxygen or nitrogen; it probably consisted mostly of water vapor plus some hydrogen and hydrogen-rich gases such as methane, hydrogen sulfide, and ammonia. The fermentation process yields ethanol and carbon dioxide. The system could have continued only until the fermentation organisms used up the original stock of "food" molecules or choked on the carbon dioxide buildup. The system stabilized temporarily with the appearance of a new organism (blue-green algae, or cyanobacteria) capable of recycling carbon dioxide into sugars and cellulose, thus again closing the carbon cycle. This new process was anaerobic photosynthesis.

However, the photosynthesis process also had a waste product: namely, oxygen. For a long time (over a billion years) the oxygen generated by anaerobic photosynthesis was captured by dissolved ferrous iron molecules, and sequestered as insoluble ferric oxide or magnetite, with the help of another primitive organism, the stromatolites. The resulting insoluble iron oxide was precipitated on the ocean bottoms.[3] (The result is the large deposits of high-grade iron ore we exploit today.) The system was still unstable at this point. It was only the evolutionary invention of two more biological processes, aerobic respiration and aerobic photosynthesis, that closed the oxygen cycle as well. Still other biological processes—nitrification and denitrification, for instance—had to appear to close the nitrogen cycle and others.

Evidently biological evolution responded to inherently unstable situations (open cycles) by "inventing" new processes (organisms) to stabilize the system by closing the cycles. This self-organizing capability is the essence of what has been called Gaia. However, the instabilities in question were slow to develop, and the evolutionary responses were also slow to evolve. It took several billion years before the biosphere reached its present degree of stability.

In the case of the industrial system, the time scales have been drastically shortened. Human activity already dominates natural processes in many respects. While cumulative anthropogenic changes to most natural nutrient stocks still remain fairly small, in most cases,[4] the *rate* of nutrient mobilization by human industrial activity is already comparable to the natural rate in many cases. Table 1 shows the natural and anthropogenic mobilization (flow) rates for the four major biological nutrients, carbon, nitrogen, phosphorus, and sulfur. In all cases, with the possible exception of nitrogen, the anthropogenic contributions exceed the natural flows by a considerable margin. The same is true for most of the toxic heavy metals, shown in Table 2.

Based on relatively crude materials cycle analyses, at least, it would appear that industrialization has already drastically disturbed, and therefore destabilized, the natural system.

TABLE 1 Anthropogenic Nutrient Fluxes, Teragrams/year (Tg/yr)

	Carbon		Nitrogen		Sulfur		Phosphorus	
	Tg/yr	%	Tg/yr	%	Tg/yr	%	Tg/yr	%
To atmosphere, total	7,900	4	55.0	12.5	93	65.5	1.5	12.5
Fossil fuel combustion and smelting	6,400		45.0		92			
Land clearing, deforestation	1,500		2.6		1		1.5	
Fertilizer volatilization[a]			7.5					
To soil, total			112.5	21	73.3	23.4	15	7.4
Fertilization			67.5		4.0		15	
Waste disposal[b]			5.0		21.0			
Anthropogenic acid deposition			30.0		48.3			
Anthropogenic (NH_3, NH_4) deposition			10.0					
To rivers and oceans, total			72.5	25	52.5	21	5	10.3
Anthropogenic acid deposition			55.0		22.5			
Waste disposal			17.5		30.0		5	

[a]Assuming 10% loss of synthetic ammonia-based fertilizers applied to land surface (75 Tg/yr).
[b]Total production (= use) less fertilizer use, allocated to landfill. The remainder is assumed to be disposed of via waterways.

TABLE 2 Worldwide Atmospheric Emissions of Trace Metals (1,000 tonnes per year)

Element	Energy Production	Smelting, Refining, and Manufacturing	Manufacturing Processes	Commercial Uses, Waste Incineration, and Transportation	Total Anthropogenic Contributions	Total Contributions by Natural Activities
Antimony	1.3	1.5	—	0.7	3.5	2.6
Arsenic	2.2	12.4	2.0	2.3	19.0	12.0
Cadmium	0.8	5.4	0.6	0.8	7.6	1.4
Chromium	12.7	—	17.0	0.8	31.0	43.0
Copper	8.0	23.6	2.0	1.6	35.0	6.1
Lead	12.7	49.1	15.7	254.9	332.0	28.0
Manganese	12.1	3.2	14.7	8.3	38.0	12.0
Mercury	2.3	0.1	—	1.2	3.6	317.0
Nickel	42.0	4.8	4.5	0.4	52.0	2.5
Selenium	3.9	2.3	—	0.1	6.3	3.0
Thallium	1.1	—	4.0	—	5.1	29.0
Tin	3.3	1.1	—	0.8	5.1	10.0
Vanadium	84.0	0.1	0.7	1.2	86.0	28.0
Zinc	16.8	72.5	33.4	9.2	132.0	45.0

NOTE: Based on relatively crude materials cycle analyses, at least, it would appear that industrialization has already drastically disturbed, and *ipso facto* destabilized, the natural system.

SOURCE: Nriagu (1990).

MEASURES OF INDUSTRIAL METABOLISM

There are only two possible long-run fates for waste materials: recycling and reuse or dissipative loss.[5] (This is a straightforward implication of the law of conservation of mass.) The more materials are recycled, the less will be dissipated into the environment, and vice versa. Dissipative losses must be made up by replacement from virgin sources. A strong implication of the analysis sketched above is that a long-term (sustainable) steady-state industrial economy would necessarily be characterized by near-total recycling of intrinsically toxic or hazardous materials, as well as a significant degree of recycling of plastics, paper, and other materials whose disposal constitutes an environmental problem. Heavy metals are among the materials that would have to be almost totally recycled to satisfy the sustainability criterion. The fraction of current metal supply needed to replace dissipative losses (i.e., production from virgin ores needed to maintain a stable level of consumption) is thus a useful surrogate measure of "distance" from a steady-state condition, that is, a condition of long-run sustainability.

Most economic analysis in regard to materials, in the past, has focused on availability. Data on several categories of reserves (economically recoverable, potential, etc.) are routinely gathered and published by the U.S. Bureau of Mines, for example. However, as is well known, such figures are a poor proxy for actual reserves. In most cases the actual reserves are much greater than the amounts actually documented. The reason, simply, is that most such data are extrapolated from test borings by mining or drilling firms. There is a well-documented tendency for firms to stop searching for new ore bodies when their existing reserves exceed 20 to 25 years' supply. Even in the case of petroleum (which has been the subject of worldwide searches for many decades) it is not possible to place much reliance on published data of this kind.[6]

However, a sustainable steady state, as discussed above, is less a question of resource availability than of recycling or reuse efficiency. As commented earlier, a good measure of unsustainability is dissipative use. This raises the distinction between *inherently dissipative* uses and uses for which the material could be recycled or reused, in principle, but is not. The latter could be termed *potentially recyclable*. Thus, there are really three important classes of materials use: (1) uses that are economically and technologically compatible with recycling under present prices and regulations; (2) uses that are not economically compatible with recycling but where recycling is technically feasible, for example, if the collection problem were solved; and (3) uses for which recycling is inherently not feasible. Admittedly there is some fuzziness in these classifications, but it should be possible for a group of international experts to arrive at some reconciliation.

Generally speaking, it is arguable that most structural metals and industrial catalysts are in the first category; other structural and packaging materials, as well as most refrigerants and solvents, fall into the second category. This leaves coatings, pigments, pesticides, herbicides, germicides, preservatives, flocculants, antifreezes, explosives, propellants, fire retardants, reagents, detergents, fertilizers, fuels, lubricants, and the like in the third category. In fact, it is easy to verify that most chemical products belong in the third category, except those physically embodied in plastics, synthetic rubber, or synthetic fibers.

From the standpoint of elements, if one traces the uses of materials from source to final sink, it can be seen that virtually all sulfur mined (or recovered from oil, gas, or metallurgical refineries) is ultimately dissipated in use (e.g., as fertilizers or pigments) or discarded, as waste acid or as ferric or calcium sulfites or sulfates. (Some of these sulfate wastes are classed as hazardous). Sulfur is mostly (75–80 percent) used, in the first place, to produce sulfuric acid, which in turn is used for many purposes. But in every chemical reaction the sulfur must be accounted for—it must go somewhere. The laws of chemistry guarantee that reactions will tend to continue either until the most stable possible compound is formed or until an insoluble solid is formed. If the sulfur is not embodied in a "useful" product, it must end up in a waste stream.

There is only one long-lived structural material embodying sulfur: plaster-of-Paris (hydrated calcium sulfate) which is normally made directly from the natural mineral gypsum. In recent years, sulfur recovered from coal-burning power plants in Germany has been converted into synthetic gypsum and used for construction. However, this potential recycling loop is currently inhibited by the very low price

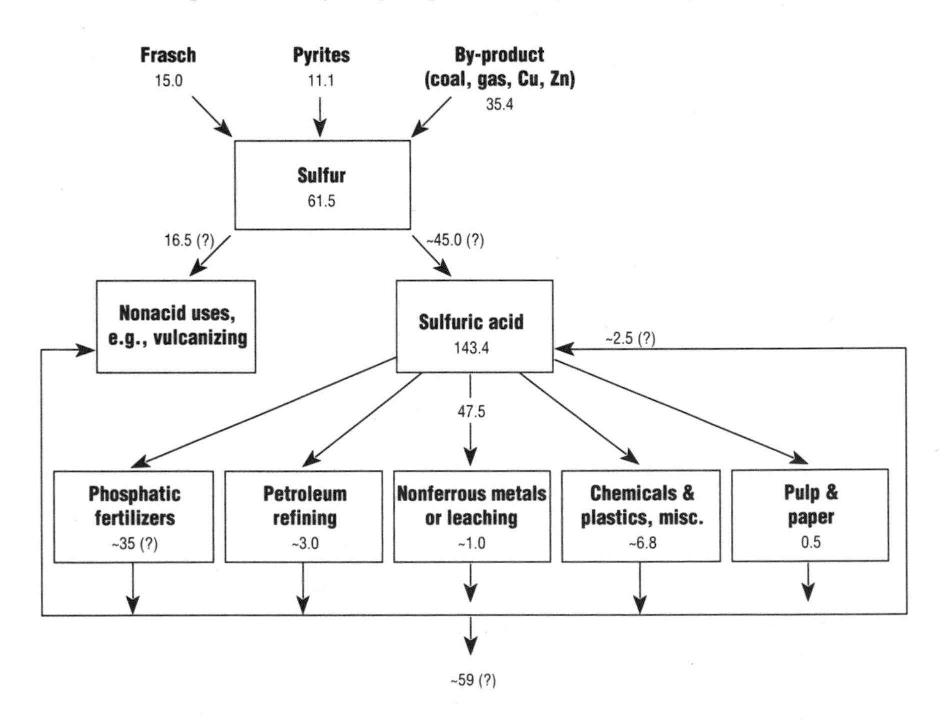

FIGURE 4 Dissipative uses of sulfur, 1988 (millions of metric tons).

TABLE 3 Examples of Dissipative Use

Substance	10^6 T	Dissipative Uses
Other chemicals		
Chlorine	25.9	Acid, bleach, water treatment, PVC solvents, pesticides, refrigerants
Sulfur	61.5	Acid (H_2SO_4), bleach, chemicals, fertilizers, rubber
Ammonia	24.0	Fertilizers, detergents, chemicals
Phosphoric acid	93.6	Fertilizers, nitric acid, chemicals (nylon, acrylics)
NaOH	35.8	Bleach, soap, chemicals
Na_2CO_3	29.9	Chemicals (glass)
Heavy metals		
Copper sulfate ($CuSO_4 \cdot 5H_2O$)	0.10	Fungicide, algicide, wood preservative, catalyst
Sodium bichromate	0.26	Chromic acid (for plating), tanning, algicide
Lead Oxides	0.24	Pigment (glass)
Lithopone (ZuS)	0.46	Pigment
Zinc Oxides	0.42	Pigment (tires)
Titanium Oxide (TiO_2)	1.90	Pigment
Tetraethyl lead	?	Gasoline additive
Arsenic	?	Wood preservative, herbicide
Mercury	?	Fungicide, catalyst

of natural gypsum. Apart from synthetic gypsum, there are no other durable materials in which sulfur is physically embodied. It follows from materials balance considerations that sulfur is entirely dissipated into the environment. Globally, about 61.5 million metric tons of sulfur as sulfur—not including gypsum—were produced in 1988. Of this, less than 2 million tons were recycled (mainly as waste sulfuric acid), as indicated schematically in Figure 4.

Very little is currently used for structural materials. Thus, most sulfur chemicals belong in class 3. Following similar logic, it is easy to see that the same is true of most chemicals derived from ammonia (fertilizers, explosives, acrylic fibers), and phosphorus (fertilizers, pesticides, detergents, fire retardants). In the case of chlorine, there is a division between class 2 (solvents, plastics, etc.) and class 3 (hydrochloric acid, chlorine used in water treatment, etc.).

Chlorofluorocarbon (CFC) refrigerants and solvents are long-lived and nonreactive. In fact, this is the reason they pose an environmental hazard. Given an appropriate system for recovering and reconditioning old refrigerators and air conditioners, the bulk of the refrigerants now in use could be recovered, for either reuse or destruction. Hence, they belong in class 2. However CFCs used for foam-blowing are not recoverable. Table 3 shows world output of a number of materials—mostly chemicals—whose uses are, for the most part, inherently dissipative (class 3). It would be possible, with some research, to devise measures of the inherently dissipative uses of each element, along the lines sketched above.

TABLE 4 Scrap Use in the United States

	Total Consumption (million short tons)			% of Total Consumption in Recycled Scrap		
Material	1977	1982	1987	1977	1982	1987
Aluminum	6.49	5.94	6.90	24.1	33.3	29.6
Copper	2.95	2.64	3.15	39.2	48.0	39.9
Lead	1.58	1.22	1.27	44.4	47.0	54.6
Nickel	0.75	0.89	1.42	55.9	45.4	45.4
Iron/steel	142.40	84.00	99.50	29.4	33.4	46.5
Zinc	1.10	0.78	1.05	20.9	24.1	17.7
Paper	60.00	61.00	76.20	24.3	24.5	25.8

SOURCE: Institute of Scrap Recycling Industries (1988).

Sustainability, in the long run, would imply that such measures decline. Currently, they are almost certainly increasing.

With regard to materials that are potentially recyclable (classes 1 and 2), the fraction actually recycled is a useful measure of the approach toward (or away from) sustainability. A reasonable proxy for this, in the case of metals, is the ratio of secondary supply to total supply of final materials: see, for example, Table 4. This table shows, incidentally, that the recycling ratio in the United States has been rising consistently in recent years only for lead and iron or steel. For lead, the ban on using tetraethyl lead as a gasoline additive (an inherently dissipative use) is entirely responsible.

Another useful measure of industrial metabolic efficiency is the economic output per unit of material input. This measure can be called *materials productivity*. It can be measured, in principle, not only for the economy as a whole, but for each sector. It can also be measured for each major "nutrient" element, such as carbon, oxygen, hydrogen, sulfur, chlorine, iron, and phosphorus. Measures of this kind for the economy as a whole are not reliable indicators of increasing technological efficiency, or progress toward long-term sustainability. The reason is that increasing efficiency—especially in rapidly developing countries—can be masked by structural changes, such as investment in heavy industry, which tend to increase the materials (and energy) intensiveness of economic activity. On the other hand, within a given sector, one would expect the efficiency of materials use—or materials productivity—to increase, in general.[7]

Useful aggregate measures of the state of the environment in relation to sustainability can be constructed from physical data that are already collected and compiled in many countries. To derive these aggregates and publish them regularly would provide policymakers with a valuable set of indicators at little cost.

It is clear that other interesting and useful measures based on physical data are also possible. Moreover, if similar data were collected and published at the sectoral level, it would be possible to undertake more ambitious engineering-economic systems analyses and forecasts—of the kind currently possible only for energy—in the entire domain of industrial metabolism.

POLICY IMPLICATIONS OF THE INDUSTRIAL METABOLISM PERSPECTIVE

It may seem odd to suggest that a mere viewpoint—in contradistinction to empirical analysis—may have policy implications. But it is perfectly possible. In fact, there are two implications that come to mind. First, the industrial metabolism perspective is essentially "holistic" in that the whole range of interactions between energy, materials, and the environment is considered together—at least, in principle. The second major implication, which virtually follows from the first, is that from this holistic perspective it is much easier to see that narrowly conceived or short-run (myopic) "quick fix" policies may be far from globally optimum. In fact, from the larger perspective, such policies can be harmful.

The best way to explain the virtues of a holistic view is by contrasting it with narrower perspectives. Consider the problem of waste disposal. It is a consequence of the law of conservation of mass that the total quantity of materials extracted from the environment will ultimately return thence as some sort of waste residuals or "garbo-junk" (Ayres and Kneese, 1969, 1989). Yet environmental protection policy has systematically ignored this fundamental reality by imposing regulations on emissions *by medium.* Typically, one legislative act mandates a bureaucracy that formulates and enforces a set of regulations dealing with emissions by "point sources" only to the air. Another act creates a bureaucracy that deals only with waterborne emissions, again by "point sources." And so forth.

Not surprisingly, one of the things that happened as a result was that some air pollution (e.g., fly ash and SO_x from fossil fuel combustion) was eliminated by converting it to another form of waste, such as a sludge to be disposed of on land. Similarly, some forms of waterborne wastes are captured and converted to sludges for land disposal (or, even, for incineration). Air and water pollution were reduced, but largely by resorting to land disposal. But landfills also *cause* water pollution (leachate), and air pollution, due to anaerobic decay processes. In short, narrowly conceived environmental policies over the past 20 years and more have largely shifted waste emissions from one form (and medium) to another, without significantly reducing the totals. In some cases, policy has encouraged changes that merely dilute the waste stream without touching its volume at all. The use of high stacks for coal-burning power plants, and the building of longer sewage pipes to carry wastes farther offshore exemplify this approach.

To be sure, these shifts may have been beneficial in the aggregate. But the costs have been quite large, and it is only too obvious that the state of the environment "in the large" is still deteriorating rapidly. One is tempted to think that a more holistic approach, from the beginning, might have achieved considerably more at considerably less cost.

In fact, there is a tendency for suboptimal choices to get "locked in" by widespread adoption. Large investments in so-called clean coal technology would surely extend the use of coal as a fuel—an eventuality highly desired by the energy establishment—but would also guarantee that larger cumulative quantities of sulfur, fly ash (with associated toxic heavy metals), and carbon dioxide would be produced. The adoption of catalytic converters for automotive engine exhaust is another case in point. This technology is surely not the final answer, since it is not effective in older vehicles. Yet it has deferred the day when internal combustion engines will eventually be replaced by some inherently cleaner automotive propulsion technology. By the time that day comes, the world's automotive fleet will be two or three times bigger than it might have been otherwise, and the cost of substitution will be enormously greater.

The implication of all these points for policymakers, of course, is that the traditional governmental division of responsibility into a large number of independent bureaucratic fiefdoms is dangerously faulty.[8] Yet the way out of this organizational impasse is far from clear. Top-down central planning has failed miserably and is unlikely to be tried again soon. On the other hand, pure "market" solutions to environmental problems are limited in cases where there is no convenient mechanism for valuation of environmental resource assets (such as beautiful scenery) or functions (such as the ultraviolet radiation protection afforded by the stratospheric ozone layer). This is primarily a problem of *indivisibility*. Indivisibility means that there is no possibility of subdividing the attribute into "parcels" suitable for physical exchange. In some cases this problem can be finessed by creating exchangeable "rights" or "permits," but the creation of a market for such instruments depends on other factors, including the existence of an effective mechanism for allocating such rights, limiting their number, and preventing poaching or illicit use of the resource.

SUMMARY

Needless to say, the policy problems have economic and sociopolitical ramifications well beyond the scope of this paper. However, as the Chinese proverb has it, the longest journey begins with a single step. Developing industrial metabolism as an analytic tool certainly represents one critical step in understanding industrial ecology systems and effecting change toward sustainability.

NOTES

1. This analogy between firms and organisms can be carried further, resulting in the notion of "industrial ecology." Just as an ecosystem is a balanced, interdependent quasi-stable community of organisms living together, so its industrial analogue may be described as a balanced, quasi-stable collection of interdependent firms belonging to the same economy. The interactions be-

tween organisms in an ecosystem range from predation and parasitism to various forms of cooperation and synergy. Much the same can be said of firms in an economy.

2. A moment's thought should convince the reader that if the stock in any compartment changes, the stock in at least one other compartment must also change.
3. Another kind of primitive marine organism apparently used hydrogen sulfide as an energy source. The sulfur, released as a waste, combined with the dissolved iron and precipitated out as iron sulfide (pyrites).
4. However, this statement is not true for greenhouse gases in the atmosphere. Already the concentration of carbon dioxide has increased 20 percent since preindustrial times, while the concentration of methane is up 50 percent. The most potent greenhouse gases of all, chlorofluorocarbons, do not exist in nature at all.
5. The special case of indefinite storage in deep underground mines, wells, or caverns, currently being considered for nuclear wastes, is not really applicable to industrial or consumer wastes except in very special and rare circumstances. Surface landfills, no matter how well designed, are hardly permanent repositories, although little consideration has been given to the long-run disposal of leachates.
6. The reserve-to-production ratio has remained close to 20 years. For example, this figure was widely published in the 1920s by Graf, as cited by Rogner (1987).
7. This need not be true for each individual element, however. A major materials substitution within a sector can result in the use of one material increasing, at the expense of others, of course. The substitution of plastics for many structural materials, or of synthetic rubber for natural rubber, would exemplify this sort of substitution. Currently, glass fibers are in the process of substituting for copper wire as the major carrier of telephonic communications.
8. The analogous problem is beginning to be recognized in the private sector, as the legacy of Frederick Taylor is finally being challenged by new managerial/organizational forms. The large U.S. firms, which adopted Taylorism first and most enthusiastically at the beginning of the twentieth century, have been the slowest to adapt themselves to the new environment of intense international competition and faster technological change.

REFERENCES

Ayres, Robert U. 1988. Self organization in biology and economics. International Journal on the Unity of the Sciences 1(3)(Fall) [also IIASA Research Report #RR-88-1, 1988].

Ayres, Robert U., and Allen V. Kneese. 1969. Production, Consumption and Externalities. American Economic Review, June [Reprinted in Benchmark Papers in Electrical Engineering and Computer Science, Daltz and Pentell, eds., Dowden, Hutchison and Ross, Stroudsberg 1974 and Bobbs-Merrill Reprint Series, N.Y.].

Ayres, Robert U., and Allen V. Kneese. 1989. Externalities: Economics and thermodynamics. In Economy and Ecology: Towards Sustainable Development, Archibugi and Nijkamp, eds. Netherlands: Kluwer Academic Publishers.

Georgescu-Roegen, Nicholas. 1971. The Entropy Law and the Economic Process. Cambridge, Mass.: Harvard University Press.

Institute of Scrap Recycling Industries. 1988. Facts—1987 Yearbook. Washington, D.C.: Institute of Scrap Recycling Industries.

Nriagu, J. O. 1990. Global metal pollution. Environment 32(7):7-32.

Rogner, Hans-Holger. 1987. Energy in the world: The present situation and future options. In Proceedings of the 17th International Congress of Refrigeration, August 24-28, 1987.

The Greening of Industrial Ecosystems. 1994.
Pp. 38–60. Washington, DC:
National Academy Press.

Energy and Industrial Ecology

HENRY R. LINDEN

This discussion of energy supply and use as a critical element of industrial ecology will focus primarily on fossil fuels, broadly classified as coal, oil, and natural gas. On the production side, the term "oil" generically refers to total crude oil, lease condensate, and natural gas liquids, and "coal" includes lignite and carbonization products. On the consumption side, the term "oil" includes liquid, solid, and normally gaseous products from the refining and processing of crude oil, lease condensate, and natural gas liquids. Most of these fuels are either burned directly or oxidized indirectly, such as the use of coal or petroleum coke in the production of metals. Only a relatively small percentage of the carbon and hydrogen content of these fuels is converted to products such as plastics, road-building materials, chemicals, fertilizers, and lubricants and is, therefore, not immediately released into the biosphere.

Several factors must be considered in assessing the future role of fossil fuels in meeting future energy needs. These include the dissipative material flows from fossil fuel use into the biosphere, the benefits and costs of energy abundance, trends in energy productivity, future energy technology and utilization trends, and the magnitude of fossil fuel reserves and resources relative to the rates of consumption for oil, natural gas, and coal. These considerations suggest that three key issues will determine the future role of fossil fuels in meeting U.S. and global energy requirements:

1. *Resource depletion*: Can the further development of the global fossil fuel resource base yield sufficient quantities of oil, natural gas, and coal to permit an orderly transition to a sustainable, non-fossil-fuel-dependent global energy sys-

tem? Moreover, can this development maintain the cost-competitiveness of these fuels (absent the imposition of large carbon taxes) and their reliability as primary energy sources for a substantial share of the useful energy services required by the various end-use sectors over a long enough period—say, 100 years?

2. *Global ecology versus human needs*: Will the benefits of continued reliance on largely fossil-fuel-based energy abundance (as the engine of human economic and social progress and the foundation of improvements in the human environment) exceed the likely costs of any associated long-term ecological consequences in terms of their impact on biodiversity and other environmental values and their impact on future human well-being?

3. *Developing versus developed world equity*: Are the industrialized countries of the Western world and the Pacific Rim willing and able to help the developing countries and the nations of the former Soviet Bloc achieve eventual parity of economic and social well-being by providing them with the technical and financial assistance required to meet their energy needs in ways that are as efficient, cost-effective, and pollution-free as possible? Doing so would avoid the creation of regional and global environmental problems that could easily overwhelm any conceivable improvements in energy supply and use practices of the advanced industrial countries.

In brief, the major findings concerning these key issues are as follows:

1. Remaining fossil fuel resources—including petroleum liquids and natural gas—could meet the major share of global energy needs at costs competitive with nonfossil-fuel options during the likely transition period to a sustainable global energy system. This would be the expected outcome of continuing advances in fossil fuel extraction, conversion, transport, and end-use technologies that are enhanced by the large price and technology elasticities of energy supply and demand, and by the efficiency gains inherent in the ongoing global trend of electrification of stationary energy uses.

2. There is ample evidence that the benefits of increased energy consumption measured in terms of *human* well-being have outweighed any quantifiable environmental costs and are likely to do so for the foreseeable future. The substitution of commercial energy commodities and energy-intensive technologies for human and animal labor and primitive renewable energy forms has been a key factor in the evolution of modern democratic societies. It has reduced the need to exploit human labor pools as evidenced by the abolition of slavery, serfdom, and child labor and, more recently, the emancipation of women (an important step in population stabilization). Increased energy use will also be essential in curing the developing world's economic and social ills, which mirror the conditions in much of the Western world before the Industrial Revolution. However, more than a trillion metric tons of additional carbon would be released into the biosphere over the next 100 years if fossil fuel use continues to be determined by least-cost energy service considerations. The long-term ecological consequences of that release

and its associated impacts on human well-being remain open questions that require early resolution.

3. It is evident that without assistance from advanced industrialized countries, the developing world's efforts to improve the economic and social well-being of its burgeoning populations will most likely rely on energy technologies and practices that are relatively inefficient and environmentally undesirable. Absent such assistance, developing countries will follow the historical pattern of growth rates of primary energy consumption that approach or even exceed the rates of economic growth. In contrast, in market economies with extensive energy supply, conversion, and distribution infrastructures and access to advanced technologies, primary energy consumption typically grows only about half as fast as gross domestic product (GDP). Moreover, the ratio of the primary energy consumption to GDP continues to decline with capital-intensive, cost-effective efficiency improvements and the ongoing electrification of energy systems.

In addition, in earlier stages of industrialization inherently capital-intensive measures to minimize environmental impacts of energy supply and use tend to be given low priority. The two most populous developing countries—India and the People's Republic of China—are largely coal-dependent, and the former Soviet Bloc has to overcome a legacy of extremely low energy productivity and gross neglect of even rudimentary environmental and public health and safety standards. This heightens the urgency of creating effective mechanisms for technical and financial assistance for the part of the world whose primary energy consumption in 2020 is expected to exceed by a considerable margin the combined consumption of North America, Western Europe, and the Pacific Rim.

DISSIPATIVE MATERIALS FLOWS FROM FOSSIL FUEL USE

Of all the dissipative materials flows into the biosphere that stem from human activities, carbon dioxide (CO_2) formed in the combustion of fossil fuels is clearly the largest by mass, with the possible exception of industrial waste. For example, the United States generates approximately 12 billion short tons of wet industrial waste annually (Allen and Jain, 1992), but only 0.2 billion short tons of municipal solid waste (Kaldjian, 1990). However, the industrial waste statistics are greatly inflated because large wastewater discharges from industrial operations are included in the count. In comparison, the author estimates that in 1991, the United States' use of 2.2 billion short tons of fossil fuels resulted in the emission of 6 billion short tons of CO_2, while global use of 10.5 billion short tons of fossil fuels in 1991 generated 26 billion short tons of CO_2. The global estimate agrees closely with a 1992 estimate by the MITRE Corporation (Gouse et al., 1992). Using the normal practice of reporting CO_2 emissions, this is equivalent to 7.1 billion short (6.4 billion metric) tons per year of carbon.[1] The source of this carbon in the form of CO_2 was global consumption in 1991 of approximately 24 billion barrels or 3.6

billion short tons of oil, 75 trillion cubic feet (Tcf) or 1.6 billion short tons of natural gas, and 5.3 billion short tons of coal, as estimated by the author from a variety of sources (Mensch, 1992; U.S. Department of Energy, 1991a, 1991b).

Fossil fuel use also generates much smaller, but still substantial amounts of "acid gases"—sulfur oxides (SO_x), nitrogen oxides (NO_x), and reactive volatile organic compounds (VOCs). In 1985 the United States emitted 23 million short tons of SO_x (reported as sulfur dioxide), 21 million short tons of NO_x (reported as nitrogen dioxide), and 22 million short tons of VOCs (Saeger et al., 1989). A more recent study reports 1990 U.S. emissions of 23.3 million short tons of SO_x as sulfur dioxide, 21.6 million short tons of NO_x as nitrogen dioxide, 20.6 million short tons of non-methane hydrocarbons, 66 million short tons of carbon monoxide, and 8.3 million short tons of total suspended particulates (Alliance to Save Energy et al., 1992). The study also reports a breakdown of these emissions from stationary and mobile fossil fuel uses, as well as from miscellaneous sources, including natural sources such as forest fires. All of these emissions in the United States are on a downward trend due to federal and state legislation and regulation.

U.S. fossil fuel use currently represents roughly 21 percent of global fossil fuel use, and U.S. commercial energy consumption 24 percent of the world total. Because U.S. and global carbon emissions bear a similar relationship, one would expect the extrapolation of U.S. data to yield global emissions of SO_x, NO_x, and VOCs on the order of 100 million short tons per year each. Global data on VOCs emissions from fossil fuel use seem to be unavailable. However, the World Energy Council recently reported 1990 global emissions of 64 million metric tons of sulfur, or 141 million short tons of SO_x as sulfur dioxide, and 24 million metric tons of nitrogen, or 87 million short tons of NO_x as nitrogen dioxide (Commission on Energy for Tomorrow's World, 1992). This compares with 1980 global emissions of 121 million short tons of SO_x and 76 million short tons of NO_x (Yokobori, 1992). These two sets of data seem reasonably consistent, considering the increase in world energy consumption over this 10-year period. They also are in the range of the amounts estimated from U.S. emissions of SO_x and NO_x, although lower figures would have been expected, given U.S. leadership in controlling these emissions from both stationary and mobile sources and the use of a larger percentage of nonpolluting fuels in the U.S. primary energy mix. Global emissions of sulfur and nitrogen compounds into the atmosphere from fossil fuel combustion *and* smelting operations reported as part of a new assessment of anthropogenic nutrient fluxes by Ayres provides another basis of comparison (Ayres, in this volume). It shows annual emissions equivalent to 92 million metric tons of sulfur and 45 million metric tons of nitrogen, or 203 million short tons of sulfur dioxide and 163 million short tons of nitrogen dioxide. Primary metals production may add 44 percent to the SO_x emissions from direct fossil fuel use reported by the World Energy Council, but seems unlikely to nearly double NO_x emissions. The Intergovernmental Panel on Climate Change reports 1990 anthropogenic sulfur emissions of 98 million metric tons (World Meteorological Organization and United Nations Environment Program, 1992), which supports the value cited by Ayres.

In addition to reactive volatile organic compounds, the energy system also emits methane, a chemically quite unreactive greenhouse gas whose direct global warming potential over a 100-year period has recently been reassessed at eleven times that of carbon dioxide (World Meteorological Organization and United Nations Environment Program, 1992).[2] Of the total estimated anthropogenic methane emissions of 506 million metric tons in 1990, approximately 100 million metric tons (or 20 percent) is believed to be of fossil fuel origin (World Meteorological Organization and United Nations Environment Program, 1992). However, other studies give a range of 440 to 640 million metric tons of annual anthropogenic methane emissions, with fossil fuels responsible for only 12 to 16 percent of the total (Alliance to Save Energy et al., 1992). Half of this is credited to emissions of methane from coal, largely from the release of methane from the coal bed during mining operations. For an excellent summary of the sources of anthropogenic methane emissions, including the relatively minor contributions from natural gas production, transmission, distribution, and use, see the Alliance to Save Energy et al. (1992). That publication also presents a comprehensive analysis of material flows—including CO_2, methane, nonmethane hydrocarbons, SO_x, NO_x, carbon monoxide, and total suspended particulates—into the atmosphere from the U.S. energy system under alternative energy future scenarios.

Heavy metals present in coal and residual fuel oil, are released in some form during combustion and enter the biosphere, although the exact pathways are not well defined. A 1990 study cited by Ayres reports that approximately 200,000 metric tons of potentially toxic heavy metals are released to the atmosphere from fossil fuels use and that these emissions make up a significant share (about 27 percent) of total anthropogenic contributions of such metals, although fossil fuels contribute relatively little to the emissions of arsenic, cadmium, zinc, and especially lead (Ayres, in this volume). However, fossil fuels contribute the major share of vanadium and nickel and a significant share of the chromium, copper, manganese, selenium, tin, mercury, and antimony. Natural contributions to atmospheric trace metals emissions are broadly of the same order of magnitude as total anthropogenic emissions, except that natural lead emissions are much lower and mercury emissions much higher.

It should be noted that emissions from the combustion of fossil fuels increase the natural supply of three of the four major biological nutrients: carbon dioxide in air or water, and soluble compounds of nitrogen and sulfur. (Fossil fuels do not contribute to the flux of phosphorus.) Therefore, they would be expected to destabilize the closed cycles the biosphere had developed before human intervention (Ayres, in this volume). Carbon dioxide is, of course, the largest of these in tonnage, although the anthropogenic contribution to the natural carbon cycle represents by far the smallest fraction of anthropogenic nutrient fluxes—at most 4 percent. This is based on a minimum estimate of 200 billion metric tons of carbon per year that cycles between the atmosphere and the terrestrial and ocean sinks, and a maximum total current annual anthropogenic contribution of 8 billion metric tons, including emissions from deforestation, other land use, and cement production (Ayres, in this volume; Commission on Energy for Tomorrow's World, 1992; Post et al., 1990; World Meteorological Organization and United Nations Environment Program, 1992).

Whereas claims of potentially detrimental effects of anthropogenic CO_2 emissions are of relatively recent origin and still the subject of intensive debate, emissions of SO_x and NO_x have been considered potential public health hazards and ecologically harmful for a long time. Because NO_x is more difficult to control than SO_x, the focus of legislation and regulation in the past has been primarily on the latter. The 1990 U.S. Clean Air Act Amendments are the latest embodiment of this approach. However, with consensus in the United States on appropriate caps for SO_x emissions, growing emphasis is being placed on NO_x control. Fortunately, unlike CO_2, SO_x and NO_x emissions are technically controllable to very low levels at substantial but manageable costs. The issue of heavy metals emissions is just beginning to surface but also should be amenable to technical solutions at acceptable economic penalties. Only the emission of CO_2 as a possible agent of climate change seems intractable, unless it can be shown that the threat has been overstated and that further enrichment of the atmosphere with CO_2 as a result of continued exploitation of global fossil fuel resources up to their full economic potential carries risks that are acceptable and justified by demonstrable benefits (Linden, 1991, 1992a, 1993; Singer, 1992).

BENEFITS OF ENERGY ABUNDANCE

The benefits of substituting commercial energy forms and energy-intensive technologies for human and animal labor and primitive renewable energy forms since the beginning of the Industrial Revolution appear to have vastly outweighed the costs, including environmental externalities (Linden, 1991, 1992b). Except in Marxist, fascist, and clerical dictatorships, every measure of human well-being has been tremendously enhanced by growing energy consumption. The large increases in life expectancy and declines in infant mortality and the general improvements in public health in the industrialized world followed the virtual elimination of hunger by energy-intensive agriculture and the creation of widely distributed wealth as a consequence of the explosive rise in labor and capital productivity.[3] Naturally, this dramatic change in the human condition over a period of less than 200 years is also a result of rapid advances in every field of science and technology, but growing energy abundance clearly played the major role.

In addition to the general contribution of primary energy and electricity consumption to such quantitative measures of economic well-being as gross domestic product per capita, the social benefits of energy use have been equally important. The abolition of slavery, serfdom, and child labor and of other forms of human exploitation was inextricably linked to the energy revolution. More recently, the emancipation of women in the Western world and now also the Pacific Rim can be traced directly to the lightening of their traditional household and family duties with energy-intensive labor-saving devices. An added benefit of energy abundance is the economic and social mobility it generates, partly because of enhanced physical mobility. This mobility, in turn, generates greater political and cultural freedom and promotes egalitarianism. The relatively high mobility of the East Germans may even have contributed to the final collapse of the Marxist dictatorships of the former Soviet Bloc. The energy-intensive communications revolution has also played an increasingly important role in loosening societal and political constraints. The United States has clearly been the leader in breaking class, sex, and caste barriers, consistent with its often maligned ranking as the world's most energy-intensive economy. Western Europe and Japan have lagged the United States in liberalizing their more highly structured societies and emancipating their women because of greater constraints on all elements of mobility and communications access, but they are rapidly catching up. The stress on energy abundance as the driving force for the unprecedented social advances of the past two centuries is not intended to minimize the equally important philosophical and spiritual contributions of the political, social, and religious revolutions of the same period.

It should, however, be noted that *primary* energy consumption per capita is a relatively rough indicator of human well-being. The *productivity* of primary energy consumption provides a better gauge. Primary energy consumption per unit of GDP expressed in 1980 U.S. dollars of the United States and Canada—the North American members of the Organization for Economic Cooperation and Development (OECD)[4]—is about twice that of its other members (U.S. Department of Energy, 1991b). In the so-called centrally planned economies (i.e., Marxist dictatorships), primary energy consumption per unit of GDP was four times that of the members of OECD other than the United States and Canada before the breakup of the Soviet Bloc. Primary energy consumption per capita in the United States is also roughly double that of Western Europe and Japan. There are obvious flaws in these comparisons of the energy intensities of various economies and, especially, of the former and present Marxist dictatorships with their totally artificial currency valuations. Monetary exchange rates in the market economies also deviate significantly from purchasing power parity, with the U.S. dollar generally undervalued—in some instances by a factor of two. This inflates the true energy intensity of the U.S. economy and creates the appearance of relatively equivalent economic well-being in the United States, Western Europe, and Japan in spite of substantial differences in more specific measures of affluence than GDP per capita. Moreover, the standard measures of human well-being fail to capture many of the widely recognized social benefits of the traditional U.S. policy of cheap and abundant energy.

Aside from these considerations that put in question the presumption that the

United States still "wastes" large amounts of energy, there are a number of additional, readily quantifiable factors that explain the high apparent energy intensity of the U.S. economy in relation to Western Europe and Japan. These factors include a much lower U.S. population density; extensive sub- and ex-urbanization, which results in greater cost-effectiveness and convenience and therefore wider use of personal transport than public transportation; greater free-standing home ownership and much larger living space per capita; more extreme climate variations and resulting greater use of year-round indoor climate control; and the relatively large contributions of inherently energy-intensive extractive and basic manufacturing industries to U.S. economic output. All of these considerations are relevant to judging the comparative merits of U.S. energy practices.

Without them, Japan can rightly claim that as the most energy-efficient and least CO_2-emitting of all the major industrial economies, it is the appropriate model. However, as of this writing, the yen is grossly overvalued relative to the dollar. Moreover, the Japanese economy is inherently more energy-efficient because of the demographic factors listed above and has been systematically restructured to concentrate on high-value-added, low-energy-intensity sectors. For example, the primary aluminum industry in Japan has been shut down. As a result, Japan imports large amounts of energy embedded in aluminum ingots, which does not show up in domestic energy consumption and CO_2 emission statistics, and a greal deal of pollution is exported. In addition, it is, of course, well known that the average standard of living in Japan is well below that of the United States and several other members of OECD.

TRENDS IN ENERGY PRODUCTIVITY

There is little question that technology advances will continue to reduce the relative growth rates of primary energy consumption in mature industrial societies. This will occur through cost-effective efficiency improvements paced by the ongoing electrification of many stationary energy uses. Primary energy consumption in OECD countries is now projected to grow at no more than half the rate of GDP even if energy prices remain relatively stable in deflated terms. However, electricity consumption is likely to continue to be closely linked to economic growth (Linden, 1988). Obviously, if prices escalate more sharply than expected because demand for OPEC oil production will increase to 40 million barrels per day by 2010 or even sooner, or because steep carbon taxes are imposed on fossil fuel use, the growth of primary energy demand relative to GDP may be less or might again turn negative as it did as a result of the oil price shocks of 1973/74 and 1978/79. This is not necessarily a beneficial outcome. As we learned during the 1970s and early 1980s, the initial economic impact of large energy price increases is severe (Linden, 1991). Similarly, most studies of the impact of carbon taxes to stabilize CO_2 emissions from fossil fuel use at 1990 levels or reduce them by 20 percent show a reduction of economic output by several percentage points lasting through most of the twenty-first century (Linden, 1991; Manne and Richels, 1992), although this result is sensitive to assumptions about revenue neutrality (i.e., corresponding reductions in other taxes).

The detrimental economic and social impacts of high energy prices and energy scarcity on developing countries are disproportionately severe because primary energy consumption per unit of GDP is relatively high and inelastic. The basic reasons for the inefficient use of energy in developing countries are the widespread use of energy price subsidies; inadequate energy supply and distribution infrastructures and associated low levels of electrification because of lack of capital and, especially, hard currencies; and limited access to the most desirable fuels as well as advanced conversion and end use technologies. As a result, consumption of commercial primary energy sources initially tends to increase *faster* than GDP and remains at relatively high levels relative to GDP until considerable industrialization is achieved (U.S. Department of Energy, 1987). Of course, exactly the same happened in the earlier stages of development of the currently industrialized economies. Primary energy consumption (excluding fuel wood and other primitive renewable energy forms) per constant dollar gross national product (GNP) in the United States rose until just after World War I, but has since dropped to less than half its peak value (Linden, 1988). At the same time, the share of primary energy consumption used in power generation rose from 10 to 36 percent and the efficiency of power generation improved threefold.

In contrast to the situation in developing countries, energy consumption in industrial countries is not only highly technology elastic but also extremely price elastic. Because of the long history of relatively high energy costs in Western Europe and Japan, their energy systems have evolved in an economic climate that generally favored capital investment over energy consumption. In the United States, the rational economic decision was to favor energy consumption over capital investment in more efficient energy systems until the price shocks beginning with the 1973/74 oil embargo attempt. However, as a result of multiple price shocks, U.S. primary energy consumption actually contracted from its interim peaks in 1973 and 1979 and did not exceed its 1979 peak until 1988, although vigorous economic growth resumed in the early 1980s (U.S. Department of Energy, 1992).

It is unlikely that substantial differences in energy productivity between industrialized countries can persist in a globalized economy, with its highly efficient capital and energy markets and free flow of technology. This will result in growing equalization of the structure of the energy economies of major trading partners within the constraints imposed by the demographic and infrastructure limitations noted before and the differences in indigenous energy resource endowments. The pressures of domestic and world competition will continue to push each of the national energy systems toward cost-effective efficiency improvements that conform to least-cost energy service strategies—that is, the delivery of heating, lighting, cooling, refrigeration, shaft horsepower, and passenger- and ton-miles, to the end users at least cost, including the costs and benefits of *quantifiable* externalities. Even in the absence of command-and-control policies, therefore, it is likely that emissions of greenhouse gases will be reduced through market forces.

This trend toward lower greenhouse gas emissions will be enhanced as corporations and other entities take prudent measures and internalize potential future exposure to carbon taxes in their strategies and investment decisions. Some corporations already invest in preserving tropical rain forests and other means of offsetting CO_2 emission on a voluntary basis. Premature regulatory and legislative intervention based on incomplete or erroneous information, especially on an international scale, may generate fewer benefits relative to costs than these market-driven actions. Market solutions are inherently more flexible than command-and-control policies, unless prior government intervention has excessively distorted the risks and rewards of rational economic behavior. Naturally, where market imperfections inhibit pursuit of least-cost energy strategies, government can play a constructive role in removing or mitigating them. Government must also play a major role in investing in research to provide the information base for rational market behavior. This applies particularly to the issue of anthropogenic climate change and its potential costs and benefits.

OPPOSITION TO ENERGY ABUNDANCE

A number of groups in the industrialized world have had a long-standing inclination to oppose energy abundance and energy-intensive economic and social development. These views about energy abundance may have been justified when the threat of imminent depletion of the most desirable fossil fuels, namely, oil and natural gas seemed probable. The issue of intergenerational fairness was raised in this connection—the questionable morality of robbing future generations of their share of the limited resource endowment by profligate use of energy by the present generation. "Saving" energy was given special moral status over "saving" capital, labor, and other goods and services. Thus, for example, we have an "Alliance to Save Energy" but no equivalent organizations to reduce other forms of consumption. The somewhat dubious "energy crisis" further enhanced efforts to conserve energy and mobilized other important constituencies concerned about "energy security." These concerns about energy security and the quest for "energy independence" have been a long-standing U.S. preoccupation, and were so before they were heightened in the public eye by the loss of oil self-sufficiency in the 1960s and the attempts in the 1970s by the major oil-exporting countries to restrict world oil supplies for political and economic reasons (Linden, 1987-88, 1991, 1992b). In any event, the concept of self-sufficiency in oil or other critical commodities has become an anachronism in a globalized economy. This was already apparent early on in the "energy crisis" when the International Energy Agency was formed to administer oil-sharing agreements among participating industrial powers. Nothing has ever come of it or from the huge investment in the

U.S. Strategic Petroleum Reserve to moderate sharp price increases during the several instances of temporary world oil supply reductions of a few million barrels per day since the 1973/74 oil embargo attempt (Linden, 1987-88, 1991, 1992b). The benefits of unified action by major oil importers and exporters in today's interdependent world economy versus an independent U.S. pursuit of energy security were forcefully demonstrated following Iraq's invasion of Kuwait in 1990.

The historical opposition to energy abundance, and campaigns against unconstrained use of various fossil fuels (and also against nuclear power for a different set of reasons) however, did generate substantial benefits. The technical community, after premature and highly diversionary efforts to develop processes for conversion of the abundant U.S. and global coal and oil shale resources into what turned out to be uneconomic and unneeded substitutes for oil and natural gas, discovered price and technology elasticity of energy supply and demand. It initially focused on cost-effective end-use efficiency improvements with spectacular results. It then concentrated on the supply side with equally spectacular results. It soon became evident that technology advances were likely to defer the threat of oil and natural gas depletion and resulting sharp price escalations for many decades. As the threat of depletion has diminished, the focus of those opposing cheap and abundant energy has shifted to environmental issues. The newest and most effective of these issues—the threat of global warming caused by the continued unconstrained use of fossil fuels—allows opponents of energy abundance to combine saving energy with saving the planet. As noted before, most of the environmental impacts of fossil fuel use are being successfully addressed by technological development and regulation. However, other than the presently discredited nuclear option, there are no technically and economically feasible near-term responses to the global warming threat—if it is indeed a threat (Linden, 1993; Singer, 1992).

FUTURE ENERGY TECHNOLOGY AND UTILIZATION TRENDS

Over the longer term, there is a rather broad consensus among technologists that the global energy system will move toward essentially complete electrification of all stationary energy uses and that most mobile energy requirements that cannot be economically or technically satisfied with electric storage batteries will be met with hydrogen in compressed, absorbed, or liquefied form. Hydrogen for this purpose and for energy storage would be produced by water decomposition with nuclear heat, by electrolysis of water with off-peak nuclear power and wind and photovoltaic power, or by various solar heat and photochemical cycles. High-tech solutions such as geosynchronous satellites that collect and convert solar energy at high efficiency and beam it to Earth in some acceptable wave form cannot be ruled out in spite of formidable environmental problems. However, least-cost considerations suggest that this electrification trend will also require a major worldwide revival of the nuclear fission option. As a prerequisite, new reactor designs would have to incorporate such features as passive cooling that would make them inherently safe, and the fuel cycle would have to be closed through plutonium recycle or through new fuel element technologies that allow substantial burn-up of heavy fission products in the reactor. Of course, controlled thermonuclear fusion always looms on the horizon as a possible ideal alternative for creating essentially perpetual energy abundance with minimal environmental impact, but both the technological and the economic barriers still look forbidding. Even if energy systems based on nuclear fission or fusion are ruled out because of political, economic, safety, or environmental considerations, technology is on the horizon for an electricity/hydrogen energy system that uses only solar or solar-derivative energy (such as wind and ocean thermal gradient power) as input. In such a system, it is even conceivable that the atmosphere and other sinks could be "mined" for CO_2 to produce methane by reacting it with water photochemically, or to produce a wider range of organic compounds using hydrogen generated photochemically or by solar-thermal cycles.

Whatever the eventual path, time scale, and end point of this energy technology evolution turns out to be, it is clear that continued electrification will inherently improve the efficiency of the energy system and reduce its environmental impact. Electrification will also facilitate the transition from fossil fuels to more abundant or inexhaustible energy forms. However, the rationale for accelerating this process beyond the rate determined by market forces and technology advances within reasonable environmental constraints does not exist at present. Projections of catastrophic global warming before the end of the twenty-first century as a result of anthropogenic greenhouse emissions are losing credibility as the underlying science evolves (Linden, 1993; Singer, 1992). However, there is ample common ground between those who continue to believe that prudent planning requires internalization of the potential cost of some degree of global warming in energy decision making, and those who do not hold that such a view is necessary to manage the evolution of the global energy system. An acceptable compromise might be achieved by using appropriate tax and regulatory incentives to modestly accelerate the existing decarbonization and electrification trends occurring in the global energy system, within an overall conceptual framework of fully internalized, least-cost energy-service strategies.

Unfortunately, least-cost energy service strategies and market forces generally cannot be relied on to govern energy investments in the developing world. There, actions are dominated by the need to improve the quality of life of rapidly growing and often desperate populations racked by abysmal poverty, hunger, and excessive morbidity and mortality rates. Such conditions do not allow the luxury of making energy and land use decisions that conform with OECD or European Community standards. To varying degrees this observation also applies to the former centrally planned economies and those still governed by Marxist dictatorships. This is unfortunate in view of the environmental havoc they have already created by their energy, industrial, and agricultural policies. The People's Republic of China is a case in point as the world's largest coal producer and consumer of more than one billion short tons per year. China is expected to double its production and consumption by 2010. In doing so, China will overwhelm any likely CO_2 emission reductions by the industrial countries. Non-OPEC developing countries will also increase coal consumption by 200 to 300 million tons. India will be a leading contributor to this trend. More generally, primary energy consumption of the OECD in the absence of carbon taxes or excessive oil prices is projected to increase from about 180 quadrillion Btu (quads)[5] to 230 quads by 2010 (U.S. Department of Energy, 1991b). Even if the OECD decides to constrain fossil fuel use and energy use through taxes or other measures, the rest of the world is unlikely to follow suit and is expected to increase its consumption from 170 quads to 230 quads by 2010 (U.S. Department of Energy, 1991b). Most of this energy demand growth will be met with fossil fuels since nuclear power has temporarily lost its competitive edge and public acceptance in many parts of the world, and the high capital costs of wind, photovoltaic, and hydropower and other renewable energy resources will continue to impede their deployment. In the developing world, this fossil fuel use will generally be in inefficient and highly polluting small installations and appliances because electricity and gas transmission and distribution facilities are inadequate.

Over the longer range, it is clear that energy extraction, conversion, and end-use technologies will have to continue to improve at a rapid pace to meet the needs of a global population that the World Bank estimates will exceed 11 billion in the year 2100. At that time, primary energy requirements are estimated at 900 to 1,500 quads after reaching 600 to 900 quads in 2060 (Gouse et al., 1992), compared with 350 quads today. Fortunately, as will be shown below, the economically recoverable fossil fuel resource base is sufficient to allow for an orderly transition to a sustainable global energy system over this period—a system that will, of necessity, depend increasingly on renewable and nuclear fission or fusion options and be largely electrified.

FOSSIL FUEL RESERVES, RESOURCES, AND CONSUMPTION

Proved global crude oil and natural gas liquids reserves are now roughly 1.1 trillion barrels, corresponding to more than 45 years of supply at current rates of consumption.[6] Thus, proved reserves have increased by 400 billion barrels since the days of the "energy crisis" while another 400 billion barrels of crude oil were used over that period. At the same time, estimates of remaining ultimately (i.e., technically) recoverable oil and natural gas liquids resources have increased to 2 trillion barrels. A major part of this turnaround in world oil supply prospects has been due to advances in exploration, development, and production technologies—advances that continue at a rapid pace. It is, therefore, not improbable that today's 2 trillion barrels of remaining oil potential will eventually turn out to be not only technically but also economically recoverable. This would meet current rates of consumption for 80 years. The true global oil potential is likely to be higher—there has been limited exploration of so many frontier areas, and huge amounts of more marginal liquid fuel resources, such as oil shale, tar sands, and other bitumens, may be economically exploited in the light of ongoing technology improvements. It is entirely feasible to convert a major share of these oil resources into transportation fuels that meet very stringent emission standards at costs that compare favorably with many other proposed solutions.

Even in the United States, there remains a large potential for cost-effective recovery of crude oil from the remaining resource base. In contrast to year-end 1992 proved reserves of only 24 billion barrels, a recent (October 1992) authoritative study conducted under the auspices of the U.S. Department of Energy projects the following remaining recoverable resources at various price (in 1992 dollars) and technology assumptions (Fisher et al., 1992):

Existing technology, $20/barrel	99 billion barrels
Advanced technology, $20/barrel	142 billion barrels
Existing technology, $27/barrel	130 billion barrels
Advanced technology, $27/barrel	204 billion barrels

A 1989 study by the American Association of Petroleum Geologists reported in Fisher et al. (1992) arrived at a very similar result: an estimate of 247 billion barrels of crude oil recoverable as of December 31, 1986, with advanced technology at prices of $25 to $50 per barrel (1986 dollars). Although no comparable data on price and technology elasticity are available, assessments of the remaining U.S. liquid fuels resource potential must also consider remaining technically recoverable natural gas liquids. A somewhat out-of-date estimate is 30 to 35 billion barrels (Institute of Gas Technology, 1989), compared to current proved reserves of 7.5 billion barrels. Thus, while it is much more profitable and practical to explore for and produce oil overseas because of the relatively high costs and growing regulatory constraints in the United States, the United States could theoretically meet its current liquid fuels requirements of 6 billion barrels per year from conventional domestic resources at a relatively modest cost penalty for as long as 40 years. The remaining crude oil resources recoverable at up to $27 per barrel with advanced technology are equivalent to 75 years of supply at the 1991 production rate of 2.7 billion barrels. In addition, advances in an environmentally benign, in-situ oil shale retorting technology that extracts high-quality shale oil by electric radio-frequency heating to 650–725°F, could make up to 400 billion barrels of shale oil recoverable at $20 to $30 per barrel from the richest portion of the western oil shale deposits—the subsection of the Piceance Basin that assays at least 25 gallons per ton of oil shale (Bridges and Sresty, 1991; Bridges and Streeter, 1991).

The global and U.S. natural gas outlook is even more reassuring than the outlook for liquid fuels. Proved global natural gas reserves are 4,400 to 4,900 trillion cubic feet (Tcf), and conservative estimates of remaining ultimately recoverable resources are in the range of 8,000 to 9,000 Tcf, compared with annual use of 75 Tcf—that is, a 60- to 120-year supply at current rates of consumption. These gas reserves and resources represent two-thirds to three-quarters of the heating value of oil reserves and resources. This undoubtedly understates the true global natural gas potential by a wide margin on the basis of U.S. experience. Technology advances have already made huge quantities of so-called unconventional gas resources—gas from "tight" sandstone and Devonian shale reservoirs and from coal beds—economically recoverable. In addition, better reservoir characterization techniques, and improved exploration, development, production, and well stimulation technologies have vastly increased projected recoveries from conventional formations. As a result, in contrast with U.S. lower-48 proved reserves of only 155 Tcf at year-end 1992, a number of authoritative sources, including the National Petroleum Council and the U.S. Department of Energy, now project that the remaining lower-48 natural gas potential is 1,000 to 1,600 Tcf, with 1,300 Tcf a reasonable consensus value. This is equivalent to 50 to 80 years of supply at current rates of consumption without taking credit for the growing imports from Canada and the potential lower-48 use of the large Alaskan resources. Based on studies by the Gas Research Institute (Woods, 1991) and the National Petroleum Council (Natural Gas Week, 1992), at least half of these lower-48 natural gas resources are currently recoverable at less than $4 to $5 per million Btu in constant dollars, and, with continuing extraction technology advances, at less than $3 to $3.50 per million Btu.[7] Without allowance for the higher transportation and storage costs for natural gas, this would be equivalent in heating value to $17 to $30 per barrel of oil.

If the U.S. experience is extended to a global basis, recovery of as much as 10,000 Tcf at acceptable costs seems achievable—enough to meet 40 percent (rather than today's 21 percent) of current world primary energy consumption with natural gas for 70 years. Some global energy futures scenarios now assume ultimate recoveries as high as 12,000 Tcf (World Meteorological Organization and United Nations Environment Program, 1992). Natural gas causes little pollution (essentially no SO_x and readily controllable NO_x emissions), and causes total emissions from wellhead to point of use of only one-half of the CO_2-equivalent greenhouse gases caused by conventional coal-fired power generation and associated coal-mining operations. In compressed or liquefied form, natural gas is also the most promising option among alternative transportation fuels. Therefore, stress on global natural gas resource development and use would provide an obviously ideal complement to electrification. Continued direct use of natural gas in many stationary applications, compressed natural gas for automotive fleet use, and gas-fired electric power generation with combined-cycle combustion/steam turbine systems is also consistent with least-cost energy service criteria over the foreseeable future.

Coal is, of course, by far the most abundant global fossil fuel resource. Proved reserves of all ranks of coal, including lignite, are in the 1.1 to 1.8 trillion short ton range, and remaining ultimately recoverable resources are on the order of 7 trillion short tons. At current rates of consumption, proved reserves alone represent 300 years of supply. The United States has the world's largest endowment in coal and lignite resources—260 billion short tons of proved reserves and 1.0 to 1.8 trillion short tons remaining ultimately recoverable resources, compared with annual production of roughly 1 billion short tons. From a fully internalized least-cost energy service perspective, it may be premature to conclude that coal, because of its often high sulfur content and high CO_2 emissions, has an inherent disadvantage in meeting future energy requirements. Tremendous technological advances in clean and more efficient coal use have been made and continue at a rapid pace. Typical examples are atmospheric and pressurized fluidized-bed boilers in which limestone removes most of the sulfur and lower combustion temperatures reduce NO_x emissions to acceptably low levels. The latest embodiments of integrated coal gasification and combined-combustion/steam turbine systems are not only inherently capable of reducing SO_x and NO_x emissions to very low levels, they also increase the thermal efficiency of power generation from 30–33 percent to 40–43 percent and should eventually be able to achieve efficiencies in excess of 45 percent through further improvements in power generation train performance and plant design (Hertz et al., 1992; Holt, 1991). The first-cost premiums of these high-efficiency clean coal technologies over conventional power generation systems are rapidly declining and may soon disappear as their commercialization proceeds.

In aggregate, the energy content of the remaining technically recoverable conventional fossil fuels is at least 160,000 quads, or more than 450 years of potential supply at current rates of global consumption. If one includes the technically recoverable energy content of all remaining fossil fuels, including oil shale, tar sands and other bitumens, and unconventional sources of natural gas, global resources are on the order of 200,000 quads, equivalent to almost 600 years of consumption at current rates. Clearly, in view of the demonstrated price and technology elasticity of the supply and demand of fossil fuels, resource depletion is not an issue, even if between now and the year 2100 primary energy consumption trends follow the highest fossil fuel use scenarios of energy futures studies of the MITRE Corporation and the Intergovernmental Panel on Global Climate Change (Gouse et al., 1992; World Meteorological Organization and United Nations Environment Program, 1992). However, the potential ecological impact of liberating 700 billion to 2 trillion metric tons (depending on the projected increase in primary energy consumption and the projected decline in its carbon intensity) of the 4 to 5 trillion metric tons of carbon contained in these huge technically recoverable fossil fuel resources quite obviously is an issue.

FOSSIL FUELS IN THE EVOLUTION OF THE ENERGY SECTOR

The following snapshot of the world energy system reflects 1990/91 data:

- World energy production and consumption are about 350 quadrillion Btu (quads) annually and are expected to grow to 460 quads by 2010.
- Of the present consumption, 140 quads, or 65 million barrels per day or 24 billion barrels per year, is derived from liquid hydrocarbons; 75 quads or 75 trillion cubic feet per year from dry natural gas; 95 quads or 5.3 billion short tons per year from coal; 20 quads from nuclear energy; and 25 quads from hydropower and other renewables. (Use of nuclear energy and hydropower is reported in terms of actual or fossil-fuel-equivalent heat rates in power generation and, due to rounding to the nearest 5 quads, the total adds up to 355 quads.)
- Electricity production is 11 trillion kilowatt-hours annually, derived from 2.6 billion kilowatts of installed capacity, of which 315 million kilowatts (or 12 percent) is nuclear. Roughly one-third of primary energy consumption is for power production globally, compared with about 36 percent in the United States.
- Annual carbon emissions from fossil fuel use are in excess of 6 billion metric tons; SO_x and NO_x emissions are roughly 130 and 80 million metric tons, respectively; methane emissions are 55 to 100 million metric tons; and heavy metals emissions are about 200,000 metric tons.
- Electricity consumption, because ongoing electrification tends to offset efficiency gains, continues to increase roughly in proportion to national and global economic output. Primary energy consumption in mature industrial economies (including the United States and the other members of OECD) increases at no more than one-half the rate of GDP, whereas in developing economies it increases more rapidly—sometimes as fast or even faster than GDP.
- Energy consumption per unit of economic output (GDP) in the United States and Canada is about twice that of the other members of OECD because of a variety of quantifiable factors that can be readily differentiated from "waste." On the other hand, energy consumption per unit of GDP in the former Soviet Union and its Eastern European satellites was four times that of Western Europe and Japan and twice that of the United States, and is still roughly at that level in the People's Republic of China.

In considering the further evolution of this energy system, one must first decide if more CO_2 is good, bad, or indifferent for the environment. If CO_2 emissions (and fossil-fuel-related methane emissions) pose no imminent threat, then there are only two limitations on the use of fossil fuels: the detrimental economic and social consequences of excessive rates of depletion and the need to minimize potentially harmful SO_x, NO_x, and reactive volatile organic compound emissions to the environment (although SO_x and NO_x, like CO_2, are part of biospheric nutrient fluxes). Environmental considerations would also require some degree of control of heavy metals emissions from coal and residual fuel oil combustion.

Resource depletion is not a critical issue for the foreseeable future, if the world community wishes to maintain an unimpeded flow of energy commodities. Additions to known global oil and natural gas reserves continue to outstrip consumption by a wide margin, thanks largely to technology advances. This has invalidated projections of inevitable shortages and huge price increases. Rapid advances in clean and more efficient coal technologies that can substitute for oil and gas use in such important sectors as power generation further reduce the threat of supply inadequacies or sharp price escalations. However, even in assuming that fossil fuels will continue to meet the major share of global primary energy requirements for another 100 years because their inflation-adjusted prices will increase only moderately and no prohibitive carbon taxes will be imposed, substantial further efficiency improvements can still be justified economically. In fact, there are no longer any credible dissenters from the view that choices among options for energy supply and end use must be based on least-cost strategies to meet energy service requirements (heating, lighting, cooling, refrigeration shaft horsepower, passenger- or ton-miles, etc.). By definition, least-cost strategics mandate the incorporation of all available cost-effective efficiency improvements and the internalization of the costs of any quantifiable externalities of the various competing options to meet a given energy service demand.

On this basis and on the presumption that transition to an essentially inexhaustible and environmentally benign global energy system would take no longer than about a century, fossil fuels will most likely meet a major share of global energy needs during this period except for the potential problem of greenhouse gas emissions. As noted above, the quantities and extraction costs of remaining fossil fuel resources are unlikely to constrain such continued dependence, and there are a variety of technically and economically feasible options to control the emission of sulfur and nitrogen oxides and volatile organic compounds from the use of fossil fuels. It should also be possible to find solutions to the heavy metals emission problem in residual fuel oil and coal combustion. Continued electrification will be a critical element in this evolution to a sustainable global energy system. Electrification not only improves the efficiency and reduces the environmental impact of energy supply and end use but also facilitates the eventual transition from fossil fuels to even more abundant or inexhaustible energy forms. An important adjunct to electrification is wider use of natural gas because of its huge resource base and growing availability, low pollutant and CO_2 emissions, and economic advantages as a fuel for power generation and transportation.

It is also important that in an environmental strategy largely based on electrification, coal be allowed to continue to play a major role because of its abundance, low cost, and wide geographical distribution. Fortunately, power generation systems that will be able to produce electricity from coal at only 70 to 75 percent of the energy input and CO_2 emissions of currently available options, without SO_x, and with acceptable NO_x emissions now appear both technically and commercially feasible. These and other power generation and end use efficiency improvements will greatly reduce the rate of increase in CO_2 emissions, especially if they are also implemented by developing economies. However, if this is insufficient because CO_2 emissions from fossil fuel combustion are found to be so detrimental to the environment that they have to be capped at current levels or reduced by 20 percent or more, the next logical option would be electrification of the energy system using nuclear fission in conjunction with several reasonably cost-competitive renewable options. Obviously, the reactor designs in such an expansion of nuclear power would have to incorporate such features as passive cooling that would make them inherently safe and the fuel cycle would have to be closed even though this is not economic at present. During a 100-year transition period to a sustainable energy system, fusion power may also become practical in spite of the formidable technical and economic barriers it faces today, in addition to unresolved environmental issues. Over this time frame it is likely that nearly all stationary energy uses would be electrified and that hydrogen in liquefied, absorbed, or compressed hydrogen, hydrogen-powered fuel cells, and electric batteries would gradually displace hydrocarbon and alcohol fuels to meet transportation needs. Hydrogen would be produced by water decomposition with nuclear heat, by electrolysis of water with off-peak nuclear and hydropower, and wind and photovoltaic power, or by various solar-thermal and photochemical cycles.

CONCLUSIONS

From an industrial ecology viewpoint, it appears that the global energy system, while contributing a large share of dissipative material flows into the biosphere, also has sufficient flexibility and technology development potential to keep its environmental impact within acceptable bounds. However, it is still uncertain if by far the largest of these dissipative material flows—carbon dioxide emitted in the combustion of fossil fuels—has a net detrimental impact on the global ecosystem. The carbon contained in carbon dioxide is the source of the major biological nutrient—carbon, so that the small anthropogenic contributions to the annual biospheric carbon flux may have quantifiable beneficial effects. But carbon dioxide is also the major greenhouse gas. Substantial further increases in its atmospheric concentrations could, therefore, cause undesirable changes in global climate, even though the costs of global warming caused by continued optimal fossil fuel use over a reasonable transition period to a sustainable energy system may have been overstated, especially in light of the benefits of energy abundance. Yet, a credible case can still be made that over the long term the emissions of carbon dioxide, other greenhouse gases, sulfur oxides, nitrogen oxides, and heavy metals associated with the unconstrained exploitation of the abundant remaining resources of fossil fuels could pose a significant threat to the global ecology and human well-being. However, technology advances that accelerate existing trends in electrification, decarbonization, and efficiency gains of the global energy system, as well as more rigorous control of conventional pollutant emissions, should defer this threat sufficiently to avoid the economic and social harm inherent in many proposed policy responses.

These expectations for rational evolution of the global energy system in conformance with ecological constraints as well as economic and social objectives depend to an important extent on the ability of the developing world to meet its rapidly growing energy requirements in ways that do not excessively destabilize relatively fragile biospheric cycles on which long-term human well-being depends. This clearly requires major technological and probably financial assistance to the populous nations of the developing and the former and present communist world to ensure that their often highly coal-dependent energy systems make use of the most efficient and cost-effective and least-polluting energy supply and end use options. In any event, the task of managing the transition to a sustainable global energy system that can satisfy human needs over the indefinite future with minimal environmental impact would be greatly simplified if there were no need to curtail greenhouse emissions from fossil fuel use beyond the sizable reductions inherent in the pursuit of least-cost energy service strategies. The huge global endowment in economically recoverable natural gas, oil, and coal resources could then provide a century of lead time before alternative energy forms would have to meet the major share of global demand.

NOTES

1. The value of 6.4 billion metric tons of fossil fuel carbon emissions is slightly higher than found in some of the current literature, such as the 5.9 billion metric tons reported for 1990 by the World Energy Council (Commission on Energy for Tomorrow's World, 1992), perhaps, in part, because it includes metallurgical uses. Emissions of CO_2 from cement production and flaring of natural gas are other large anthropogenic carbon sources closely related to fossil fuel use, but 1990 data that purport to include these sources are also in the 6.0 billion metric ton range (Starr and Smith, 1992). Total global 1990 anthropogenic carbon emissions of 7.4 billion metric tons are reported by the Intergovernmental Panel on Climate Change, including energy production and use, cement production, and deforestation (World Meteorological Organization and United Nations Environment Program, 1992). Robert U. Ayres assesses total anthropogenic carbon emissions at 7.9 billion metric tons annually—6.4 billion tons from fossil fuel combustion and smelting, and 1.5 billion tons from deforestation (Ayres, in this volume). (1 metric ton equals 1.109 short tons).
2. According to the 1992 report of the Intergovernmental Panel on Climate Change (World Meteorological Organization and United Nations Environment Program, 1992), the indirect global warming potential roughly doubles this, which would be in general agreement with the total greenhouse effect of methane of 20 or 21 times the CO_2 value generally reported in the literature.
3. The World Energy Council has published excellent correlations of life expectancy, infant mortality, and even literacy with per capita energy and electricity consumption for 100 to 127 countries based on current statistics (Commission on Energy for Tomorrow's World, 1992).
4. The members of OECD are Australia, Austria, Belgium, Canada, Denmark, Finland, France, Germany, Greece, Iceland, Ireland, Italy, Japan, Luxembourg, the Netherlands, New Zealand, Norway, Portugal, Spain, Sweden, Switzerland, Turkey, the United Kingdom, and the United States.
5. 1 quad, or 1×10^{15} Btu, equals 1.055×10^{18} joules, or 1.055 exajoules.

6. The data reported in this section represent the author's synthesis of data from Gouse et al. (1992), Commission on Energy for Tomorrow's World (1992), Institute of Gas Technology (1989), Mensch (1992), and U.S. Department of Energy (1991a).
7. The gross (higher) heating value of dry natural gas is typically 1,000 to 1,050 Btu per standard cubic foot, with variations in global standards generally within 932 to 1,141 Btu per standard cubic foot (U.S. Department of Energy, 1991a), so that costs or prices per million Btu are roughly equivalent to costs and prices per 1,000 cubic feet.

REFERENCES

Allen, P. T., and R. Jain, eds. 1992. Special Issue on "Industrial Waste Generation and Management," Hazardous Waste and Hazardous Materials 9(1):1 111.

Alliance to Save Energy, American Gas Association, and Solar Energy Industries Association. 1992. An Alternative Energy Future. Catalog No. F81000. Arlington, Virginia: American Gas Association.

Bridges, J. E., and G. C. Sresty. 1991. An Update on the Radio-frequency In-situ Extraction of Shale Oil. 24th Oil Shale Symposium Record, Colorado School of Mines, Golden, Colorado, April 19, 1991.

Bridges, J. E., and W. S. Streeter. 1991. Economic Aspects for Oil Shale Production Using RF In-situ Retorting. 24th Oil Shale Symposium Record, Colorado School of Mines, Golden, Colorado, April 19, 1991.

Commission on Energy for Tomorrow's World. 1992. Draft Summary Global Report (Presented at 15th WEC Congress, Madrid, Spain, September 1992). London: World Energy Council.

Fisher, W. L., Noel Tyler, Carol L. Ruthven, Thomas E. Burchfield, and James F. Pantz. 1992. An Assessment of the Oil Resources Base of the United States. Bartlesville Project Office, U.S. Department of Energy, Bartlesville, Oklahoma, October 1992. U.S. Department of Energy Document No. DOE/BC-93/1/SP.

Gouse, S. W., D. Gray, G. C. Tomlinson, and D. L. Morrison. 1992. Potential World Development Through 2100: The Impact on Energy Demand, Resources and the Environment," The MITRE Corporation, McLean, Virginia, 1992. (Presented at 15th World Energy Congress, Madrid, Spain, September 1992.)

Hertz, N., N. Stewart, and A. Cohen. 1992. High-efficiency GCC power plants. EPRI Journal 17(July/August):39-42.

Holt, N. 1991. Highly efficient advanced cycles. EPRI Journal 16(April/May):40-43.

Institute of Gas Technology. 1989. IGT world reserves survey as of December 31, 1987. Chicago, Illinois: Institute of Gas Technology.

Kaldjian, P. 1990. Characterization of Municipal Solid Waste in the United States: 1990 Update. Report No. EPA—530/SW-90/042, June. Washington, D.C.: Office of Solid Waste and Emergency Response, U.S. Environmental Protection Agency.

Linden, H. R. 1987/1988. World oil—An essay on its spectacular 120-year rise (1859-1979), recent decline, and uncertain future. Energy Systems and Policy 11(4):251-266.

Linden, H. R. 1988. Electrification of energy supply—An essay on its history, economic impact and likely future. Energy Sources 10(December):127-149.

Linden, H. R. 1991. Energy, economic and social progress, and the environment: Inseparable issues of resource allocation. International Journal of Energy• Environment•Economics 1(1):1-12.

Linden, H. R. 1992a. E^3— Energy, Economy, and the Environment. Public Utilities Fortnightly 129(May 15):31-35.

Linden, H. R. 1992b. Some cautionary comments on reopening the energy policy debate. Energy Systems and Policy 15(1):75-83.

Linden, H. R. 1993. A dissenting view on global climate change. The Electricity Journal 6(6):62-69.

Manne, A. S., and R. G. Richels. 1992. Buying Greenhouse Insurance: The Economic Costs of CO_2 Emission Limits. Cambridge, Mass.: MIT Press.

Mensch, H. L., ed. 1992. Energy Statistics, Vol. 15, No. 1, Institute of Gas Technology, Chicago, Illinois, 1992.

Natural Gas Week. 1992. NPC says gas supply adequate but its report is challenged. Natural Gas Week 8(November 2):7-8.

Post, W. M., Tsung-Hung Peng, W. R. Emanuel, A. W. King, V. H. Dale, and D. L. DeAngelis. 1990. The global carbon cycle. American Scientist 78(July-August):310-326.

Saeger, M., et al. 1989. 1985 NAPAP Emissions Inventory. Alliance Technology Corporation, Chapel Hill, North Carolina, November 1989. EPA Document No. 600/7-89-012a.

Singer, S. F., ed. 1992. The Greenhouse Debate Continued: An Analysis and Critique of the IPCC Climate Assessment. The Science and Environmental Policy Project, ICS Press, San Francisco, California.

Starr, C., and S. P. Smith. 1992. Atmospheric CO_2 residence time. Unpublished paper. Electric Power Research Institute, Palo Alto, Calif.

U.S. Department of Energy. 1987. International Energy Outlook 1986. Energy Information Administration, Office of Energy Markets and End Use, U.S. Department of Energy, Washington, D.C., April 1987. U.S. Department of Energy Document No. DOE/EIA-0484(86).

U.S. Department of Energy. 1991a. International Energy Annual 1989. Energy Information Administration, Office of Energy Markets and End Use, U.S. Department of Energy, Washington, D.C., February 1991. U.S. Department of Energy Document No. DOE/EIA-0219(89).

U.S. Department of Energy. 1991b. International Energy Outlook 1991: A Post-War Review of Energy Markets. Energy Information Administration, Office of Energy Markets and End Use, U.S. Department of Energy, Washington, D.C., June 1991. U.S. Department of Energy Document No. DOE/EIA-0484(91).

U.S. Department of Energy. 1992. Monthly Energy Review, October 1992. Energy Information Administration, Office of Energy Markets and End Use, U.S. Department of Energy, Washington, D.C. U.S. Department of Energy Document No. DOE/EIA-0035(92/10)

Woods, T. J. 1991. The Long-Term Trends in U.S. Gas Supply and Prices: 1991 Edition of the GRI Baseline Projection of U.S. Energy Supply and Demand to 2010. Chicago, Illinois: Gas Research Institute.

World Meteorological Organization and United Nations Environment Program. 1992.

1992 IPCC Supplement: Scientific Assessment of Climate Change, Intergovernmental Panel on Climate Change, February.

Yokobori, K. 1992. Executive Director, Organizing Committee, 16th WEC Congress, Tokyo, Japan. OECD data cited in paper presented at 1992 Energy Policy Issue Forum: Environmental Policy and Governance, Aspen Institute for Humanistic Studies, Aspen, Colorado, August 1-5, 1992.

The Greening of Industrial Ecosystems. 1994.
Pp. 61–68. Washington, DC:
National Academy Press.

Input-Output Analysis and Industrial Ecology

FAYE DUCHIN

Elsewhere I have shown how input-output techniques could provide quantitative answers to the kinds of questions raised by industrial ecology (Duchin, 1992; see also Duchin, 1990); a numerical example illustrated how both money costs and levels of pollution would be affected by alternative technological or organizational decisions. Input-output techniques are the formalization of a more general theory about how a modern economy works, a theory I call structural economics to distinguish it from the contemporary mainstream, or neoclassical, economics.

Structural economics emphasizes the representation of stocks and flows, measured in physical units, as well as associated costs and prices where these are relevant. The variables representing such physical measures (like tons of steel or tons of carbon emissions), unlike variables that are essentially symbolic or index numbers, provide a direct link to technology and to the physical world with which industrial ecology is concerned. Highest importance is placed on this data base; and the subsequent analysis directly exploits its empirical content rather than relying exclusively on its formal manipulation. The data are developed using technical expertise and practical experience as well as experimental results, technical records, and accounting information.

Structural economics makes little use of idealized abstractions like an equilibrium state of the economy. The powerful concept of optimization is used for analyzing certain types of outcomes (for example, choosing the low-cost technology among several alternatives) but is not relied on as a general solution mechanism. While the computation of "local" optima is often useful, the concept of a maximum value for global "social welfare" (for example) is simply not operational. Neoclassical economics, by contrast, does rely in all instances on the combined concepts of equilibrium and optimization. This practice has the advantage of producing unique solutions to complex problems but at too high a price: it precludes the consideration of many alternatives that industrial ecologists will want to investigate, alternatives that society may in fact deem more desirable than those that may satisfy the short-term maximization of a single, vaguely defined criterion.

Structural economics is an incomplete theory; this is a difficult status for an economist to defend because practicing economists have come to depend on a complete, although not always operational, conceptual and analytic framework. We choose, however, not to use a high level of abstraction as a mechanism to "complete" the theory. Instead, our models are "open," making use of exogenous information such as technological projections that are provided by engineers and other technical experts rather than being derived through the use of mathematical equations that describe idealized economic mechanisms.

Through our work, my colleagues and I are trying to provide a systematic and operational description of structural economics as an economist's contribution to such new fields as industrial ecology, ecological economics, or the study of global change. This is necessary because I believe that economics needs to undergo, and will undergo, profound transformation not only in theory, but especially in scope and in methodology, as the next generation of economists attempts to understand and resolve environmental problems. The scope of the field is important because it delimits the questions that can be addressed, while the methodology determines the mathematics that will represent the theoretical propositions and the strategy for collecting and using information to implement the mathematical model and help interpret the numerical results.

Industrial ecologists will come from many backgrounds and most cannot afford to be intimately involved in the transformation of other fields such as economics while trying to build their own field. For example, the proliferation of terms used here (input-output analysis, structural economics, neoclassical economics, and ecological economics) is essential for the process of sorting out and building within economics but cannot be allowed to become a jargon that fragments our common efforts.[1] In view of these considerations, I will focus my remaining comments on two related areas of scope and methodology that are of great importance both for structural economics and for industrial ecology. These are (1) the questions that we need to address and (2) the role of case studies.

THE QUESTIONS TO BE ADDRESSED

My colleagues and I have recently completed a quantitative analysis of strategies for sustainable development over the next several decades for the UN Conference on Environment and Development (UNCED) using an input-output model and data base of the world economy (Duchin et al., 1994). We have focused on the use of energy to promote development and on the associated emissions of carbon dioxide and oxides of sulfur and nitrogen. Our results show that if moderate economic development objectives are achieved in the developing countries over the next several decades, the geographic locus of emissions will continue its historic shift from the rich to the poor economies while total emissions of the principal global pollutants will increase significantly. This is true even under optimistic assumptions about pollution reduction and controls through the accelerated adoption of modern, commercially proven technologies in both rich and poor countries. Tables summarizing some of these results are shown in the Appendix.

The first type of question we need to ask is, What can be done? I conclude that significant changes in production and consumption practices and technologies are required if emissions are to be held at current levels, not to mention reduced. Consequently, industrial ecology and Design for Environment (DFE) will need to go far beyond what is generally understood as pollution prevention and explore the potential for dramatic reductions in overall use of fossil fuels. This, and near-zero net emissions of long-lived toxic chemical compounds, are the objectives laid out by Robert Ayres (1991, pp. 22-23); a possible long-term energy scenario, with a thoughtful description of the transition to it, was described by Rogner (1993) at a recent meeting of mainly European and Japanese analysts. I also conclude from this work (and from a project in which we posed similar questions in the context of the Indonesian economy) that many of the challenges to DFE in the developing countries are very different from those in the developed economies and, from a global point of view, even more crucial to resolve. Many of these problems are related to water, soil, agriculture, forestry, fisheries, and wildlife rather than manufacturing; it is clear that industrial ecology cannot afford to ignore them (see Ayres, 1991, pp. 22-23).

Industrial ecology concepts provide powerful building blocks for developing global or national economic strategies. The constituency for this work is being built at the present time. Industrial ecology also has to, and can, respond to other kinds of imperatives in order to maintain the broad base of support that will be needed to ensure its success; for example, a great deal of the work will need to focus on improving the short-term profitability of the individual corporations where product development and production actually take place. In attempting to resolve both the strategic and the tactical problems associated with the choice of product and technology from both economic and environmental points of view, some industrial ecology analysts will want to collaborate intimately with input-output analysts because our models can be a powerful tool for addressing a second type of question: What are the economic and environmental implications of each alternative strategy?

Input-output models can be viewed as a very general form of benefit-cost analysis. The latter has typically been applied only to individual projects and requires many economic assumptions (about equilibrium, optimality, and "consumer surplus," for example) that the industrial ecology analyst may not wish to make. The modern dynamic input-output model can evaluate not only the costs but also the potential contributions to reducing the volume of pollution for each strategy to improve current industrial practices while capturing the simultaneous effects on many sectors of the economy over a period of several decades. In early work along these lines, input-output models have already been used to compare the economic implications of alternative technological choices facing individual sectors of the economy (see Duchin and Lange, 1992, 1993).

These models provide the framework for evaluating industrial ecology case studies in an economy-wide context. The work of industrial ecology analysts, in turn, is needed to provide the substantive input for the input-output analyses that until now have been developed with minimal benefit from this kind of technical expertise. In this way the combination of industrial ecology and structural economics can be used to provide a realistic basis for pollution reduction and private benefit, as well as to identify the need for action by other parties, including the government.[2]

CASE STUDIES

There is an enormous overlap between the technical case studies required by the DFE analyst and the input-output analyst. Common methodology needs to be developed to ensure that such studies meet the technical standards and analytic requirements of both communities. For the UNCED project mentioned above, case studies were carried out at the Institute for Economics Analysis for the following areas, which were chosen for their intensive use of energy: electricity generation, industrial energy conservation, household energy use, motor vehicle transportation, metal fabrication and processing, construction and its major material inputs, paper, and chemicals (Duchin et al., 1994, chapters 6-15). For each, we inventoried the techniques in use in each region of the world economy and those that might be adopted over the next several decades. A quantitative description of these techniques was incorporated into the data base. Unfortunately, however, these case studies are uneven in coverage and depth and depended on the technical literature more than on first-hand knowledge and experience.

The engineering community also has a conceptual framework for describing an entire production process (see, for example, Friedlander, in this volume) and criteria for selecting activities that could benefit from more intensive investigation. Often, however, these studies make minimal use of economic criteria or of any considerations outside of the industry in question. A powerful interdisciplinary research agenda can be realized if DFE case studies are carried out as part of a collaboration between structural economists and industrial ecologists.

The perspective and tools of structural economics permit us to identify areas where new case studies may be crucial from the points of view of economic development and alleviation of environmental problems, and to analyze their implications. The production and use of fuels and feedstocks of biological origin in temperate and tropical climates would be an ambitious and important case study. Another important area is the production and use of convenient, efficient equipment for public transportation. The industrial ecologist is likely to start from a different ordering of priorities. The collaboration could prove fruitful in identifying areas for action that can satisfy the requirements of a relatively broad constituency, in quantifying the description of the inputs and outputs associated with alternative processes that might be adopted, and in providing a quantitative assessment of the economic and environmental implications of moving in these directions.

NOTES

1. Of course, industrial ecology also has sorting out to do of concepts like industrial metabolism, industrial ecology, design for environment, life cycle analysis, and ecological engineering.
2. Quality of information and the uncertainty and risks associated with drawing conclusions on the basis of these kinds of estimates and projections need to be taken explicitly into account. An excellent discussion of these issues is found in (Funtowicz and Ravetz, 1990).

REFERENCES

Ayres, Robert U. 1991. Eco-restructuring: Managing the transition to an ecologically sustainable economy. Carnegie-Mellon University, June, unpublished paper.

Duchin, Faye. 1990. The conversion of biological materials and wastes to useful products. Structural Change and Economic Dynamics 1(2):243-261.

Duchin, Faye. 1992. Industrial input-output analysis: Implications for industrial ecology. Proceedings of the National Academy of Sciences 89:851-855.

Duchin, Faye, and Glenn-Marie Lange. 1992. Technological choices, prices, and their implications for the U.S. economy, 1963-2000. Economic Systems Research 4(1):53-76.

Duchin, Faye, and Glenn-Marie Lange. 1993. The choice of technology and associated changes in prices in the U.S. economy. Submitted to Structural Change and Economic Dynamics.

Duchin, Faye, and Glenn-Marie Lange, with Knut Thonstad and Annemarth Idenburg. 1994. Ecological Economics, Technology and the Future of the World Environment. New York: Oxford University Press.

Funtowicz, Silvio O., and Jerome R. Ravetz. 1990. Uncertainty and Quality in Science for Policy. The Netherlands: Kluwer Academic Publishers.

McCormick, J. 1985. Acid Earth. Washington, D.C.: International Institute for Environment.

Rogner, Hans-Holger. 1993. Global energy futures: The long-term perspective for ecorestructuring. Text of presentation at United Nations University Symposium on Eco-restructuring, Tokyo, July.

World Commission on Environment and Development. 1987. Our Common Future. New York: Oxford University Press.

APPENDIX

This appendix contains two tables of results, discussed briefly in the text of the paper, obtained in an analysis using an input-output model and data base of the world economy. Table 1 shows the emissions of carbon dioxide and oxides of sulfur and carbon, for the period 1980 through 2020, under a scenario that assumes moderate economic growth in the developing countries, modest growth in the rich countries, and the rapid adoption in all countries of technologies that reduce emissions in part by economizing on energy and materials. The scenario, described in detail in Duchin and Lange (1992), is based on the kinds of assumptions that are described in the Brundtland Report, *Our Common Future* (World Commission on Environment and Development, 1987). While the analysis was carried out on the basis of 16 geographic regions, the results in Table 1 are presented on a more aggregated basis.

Projections for the future can be compared only with projections made by others, but for past years the technical literature contains more direct measures and estimates. The sulfur emissions reported in the first table for 1980 are compared in Table 2, on a 16-region basis, with the results of many other studies that are reported in two surveys. This table illustrates two main points. First, by using the input-output case study methodology, it is possible to build a comprehensive data base from the bottom up based on technical assumptions. Only in this way are we able to arrive at an estimate for worldwide emissions of sulfur (in the last column of the table). Second, the input-output methodology is sufficiently explicit about underlying assumptions that one can analyze the reasons for major discrepancies between the World Model results and other results. The greatest discrepancies reported in Table 2 are with the official estimates of sulfur emissions in the former Soviet Union and in Eastern Europe. In the first case, an independent estimate by another researcher is very close to ours. In both cases, it is safe to conclude that the official estimates are unrealistically low.

TABLE 1 Regional Distribution of Emissions of Carbon Dioxide, Sulfur Oxides, and Nitrogen Oxides in 1980 through 2020

a. Carbon Dioxide (10^6 metric tons)

	1980	1990	2000	2010	2020
Rich, Developed Economies	0.55	0.50	0.43	0.38	0.34
Newly Industrializing Economies	0.05	0.06	0.08	0.09	0.10
Other Developing Economies	0.14	0.21	0.27	0.32	0.36
Eastern Europe and Former USSR	0.26	0.23	0.21	0.21	0.20
World Total	1.00	1.00	1.00	1.00	1.00
World Total (levels)	4,730	5,632	7,001	8,173	9,718

b. Sulfur Oxides (10^6 metric tons of SO_2 equivalent)

	1980	1990	2000	2010	2020
Rich, Developed Economies	0.41	0.36	0.31	0.28	0.25
Newly Industrializing Economies	0.04	0.06	0.09	0.10	0.13
Other Developing Economies	0.15	0.23	0.29	0.34	0.38
Eastern Europe and Former USSR	0.40	0.35	0.31	0.28	0.24
World Total	1.00	1.00	1.00	1.00	1.00
World Total (levels)	125.0	127.3	140.8	147.3	157.2

c. Nitrogen Oxides (10^6 metric tons of NO_2 equivalent)

	1980	1990	2000	2010	2020
Rich, Developed Economies	0.51	0.45	0.38	0.33	0.27
Newly Industrializing Economies	0.07	0.08	0.11	0.13	0.15
Other Developing Economies	0.14	0.21	0.27	0.31	0.36
Eastern Europe and Former USSR	0.28	0.26	0.24	0.23	0.22
World Total	1.00	1.00	1.00	1.00	1.00
World Total (levels)	69.0	78.2	96.6	113.7	135.8

NOTE: See text for a description of the underlying assumptions.

SOURCE: Duchin et al. (1994).

TABLE 2 Estimates of Sulfur Oxide Emissions from Fossil Fuel Combustion by Region in 1980 According to Three Sources

Region		(10^6 tons of SO_2)			
		UNEP/WHO	UN/ECE	Combined	World Model[a]
1	High-Income North America	27.81[b]	27.85[b]		23.97
2	Newly Industrializing Latin America	na	na		4.20
3	Other Latin America	na	na		1.00
4	High-Income Western Europe	18.24	13.52	18.44	15.80
5	Medium-Income Western Europe	4.46	3.92	5.45	6.40
6	Eastern Europe	12.57	8.83	14.07	25.70
7	Former USSR	na	12.80 25.00[c]		23.80
8	Centrally Planned Asia	14.21	na		13.66
9	Japan	1.64	1.26		1.60
10	Newly Industrializing Asia	na	na		2.00
11	Low-Income Asia	2.23	na		2.44
12	Major Oil Producers	0.20	na		0.84
13	Other Middle East and North Africa	0.29	na		0.73
14	Sub-Saharan Africa	na	na		0.33
15	Southern Africa	na	na		2.04
16	Oceania	1.48	na		1.47

NOTES: na = not available.

1. The UNEP/WHO and UN/ECE estimates often do not cover all countries in each region. The coverage of activities is not uniform. Some countries, notably the United States, include industrial emissions from sources other than combustion.

2. Three regions are covered by both sources but with significant geographic gaps. In the column called "combined" we have aggregated the emissions for countries reported in only one of the sources and the higher of the two estimates, so as to include more economic activities, for countries included in both.

[a]Corresponds to figures in the first column of panel b in Table 1.
[b]Includes emissions from activities other than combustion.
[c]Obtained from McCormick (1985).

SOURCES: Duchin et al. (1994) based on: UN Environment Program (UNEP) and World Health Organization (WHO), 1988, *Assessment of Urban Quality*, p. 90, New York: UN; UN Statistical Commission and the Economic Commission for Europe (ECE), 1987, *National Strategies and Policies for Air Pollution Abatement*, Table I-15, New York: UN; McCormick, J. 1985, *Acid Earth*. Washington, D.C.: International Institute for Environment.

The Greening of Industrial Ecosystems. 1994.
Pp. 69–89. Washington, DC:
National Academy Press.

Wastes as Raw Materials

DAVID T. ALLEN and NASRIN BEHMANESH

Postconsumer waste, industrial scrap, and unwanted by-products from manufacturing operations should not be viewed as wastes. Rather, they are raw materials that are often significantly underused. One of the research challenges of the emerging discipline of industrial ecology will be to identify productive uses for materials that are currently regarded as wastes, and one of the first steps in meeting this challenge will be to understand the nature of industrial and postconsumer wastes.

More than 12 billion tons of industrial waste (wet basis) are generated annually in the United States (Allen and Jain, 1992; U.S. Environmental Protection Agency [EPA], 1988a,b). Municipal solid waste, which includes postconsumer wastes, is generated at a rate of 0.2 billion tons per year (EPA, 1990). When these material flows are compared with the annual output of 0.3 billion tons per year of the top 50 commodity chemicals (*Chemical and Engineering News*, Junc 28, 1993), it is apparent that wastes should not be ignored as a potential resource. While these comparisons between waste mass and the mass of commodity products make apparent the magnitude of industrial wastes, considering mass flows alone can be somewhat misleading. The extent to which industrial wastes could serve as raw materials depends not only on the mass of the waste stream, but also on the concentration of resources in the wastes. As shown in Figure 1, the value of a resource is proportional to the level of dilution at which it is present in the raw material. Resources that are present at very low concentration can be recovered only at high cost, while resources present at high concentration can be recovered economically. The primary goals of this paper will be to evaluate the flow rates and concentrations of valuable resources in waste streams and to determine the

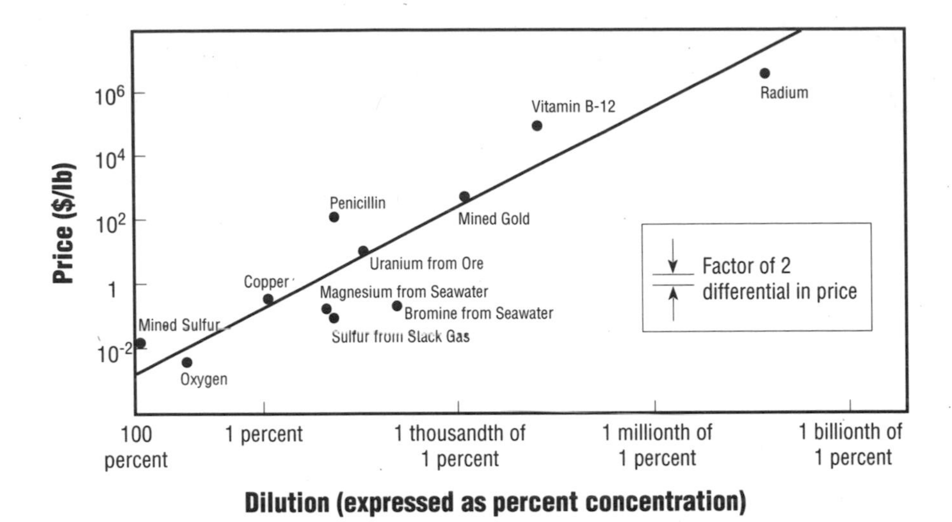

FIGURE 1 The Sherwood plot: Selling prices of materials correlate with their degree of dilution in the initial matrix from which they are being separated. Note that the horizontal axis shows increasing dilution, or decreasing concentration, in the initial matrix. SOURCE: National Research Council (1987).

extent to which materials currently regarded as wastes might be used as raw materials.

We will begin with a brief examination of the total quantities of material circulating through the waste cycles. We will then focus on a series of metals, tracking their flows as wastes and comparing the waste mass and concentration level of recycled wastes, discarded wastes, and virgin raw materials.

OVERVIEW OF INDUSTRIAL AND POSTCONSUMER WASTE GENERATION AND MANAGEMENT

Mapping the flows of more than 12 billion tons of industrial and postconsumer waste is a challenging task. Part of the challenge is integrating information from many diverse sources of data. For example, more than a dozen national sources of data on industrial wastes are available (Eisenhauer and Cordes, 1992), but each covers only a portion of industrial waste generation. Each was collected over a different time period and each considers a different subset of waste generators. Despite these difficulties, each of the data sources provides a unique perspective on industrial waste streams and can be useful. The main focus in this paper will be on just a few of these inventories, most notably the National Hazardous Waste Survey and, to a much lesser extent, the Toxic Release Inventory and the biennial survey of waste generators collected under the Resource Conservation and Recovery Act (RCRA).

The Toxic Release Inventory (TRI) (EPA, 1991), collected annually under the Superfund Amendments and Reauthorization Act (SARA), Title III, provides data on the emission profiles of more than 300 chemical species. While the TRI data are useful for profiling the releases of specific chemicals, they provide little information on the total waste stream. In contrast, the biennial survey of generators and the biennial survey of treatment, storage, and disposal facilities collected under the RCRA provide data on total waste mass but little data beyond loosely defined waste categories on the composition of the waste streams. A hybrid data base of information collected by Research Triangle Institute for the Environmental Protection Agency under RCRA (EPA, 1988c) combines some of the best features of both the TRI and the RCRA biennial survey. This data base, which is called the National Hazardous Waste Survey, combines detailed data on waste composition and data on bulk waste stream properties. It has two basic components, a generator survey focusing on waste characterization and a survey of treatment, storage, disposal, and recycling (TSDR) facilities focusing on waste treatment and disposal. The flow diagram of Figure 2 was prepared using the TSDR section of the survey and some data on air emissions from the TRI. It shows the flow patterns and approximate flow rates of industrial hazardous waste streams. All waste streams regulated under RCRA are included; also included are some wastes exempt from the RCRA regulations and some hazardous wastes managed in units exempt from RCRA permitting requirements. The total mass flow rate of all streams represented in the National Hazardous Waste Survey is approximately 0.75 billion tons per year and the total mass of releases and off-site transfers reported through the TRI totals 0.003 billion tons per year. Therefore the data represent only about 5–10 percent of the total flow rate of industrial wastes. Even though just a small fraction of industrial wastes is represented, the excluded wastes are primarily from a limited group of industries: mining, pulp and paper manufacturing, electrical power generation, and petroleum production. So, the National Hazardous Waste Survey can begin to provide a picture of the flow rates and compositions of waste streams. It is far from comprehensive and omits some major sectors of the economy that generate substantial wastes, but it represents some of the best information available on waste stream composition.

Figure 2 reports the flow rates of hazardous waste streams generated by U.S. industry in 1986, the only year for which the survey data are available. As indicated in Figure 2, a small fraction of solvent, metal, and other wastes, less than 1 percent of total waste mass generated, flows through recycling loops. The total mass involved in recycling is about 5 million tons per year (mt/yr). The largest single stream in terms of total mass flow, nearly 720 mt/yr (more than 90 percent of the total waste flow), is hazardous wastewater. Most of this stream is water with a small percentage of nonaqueous contaminants; hence the mass of the chemically hazardous component of this stream is within an order of magnitude of the

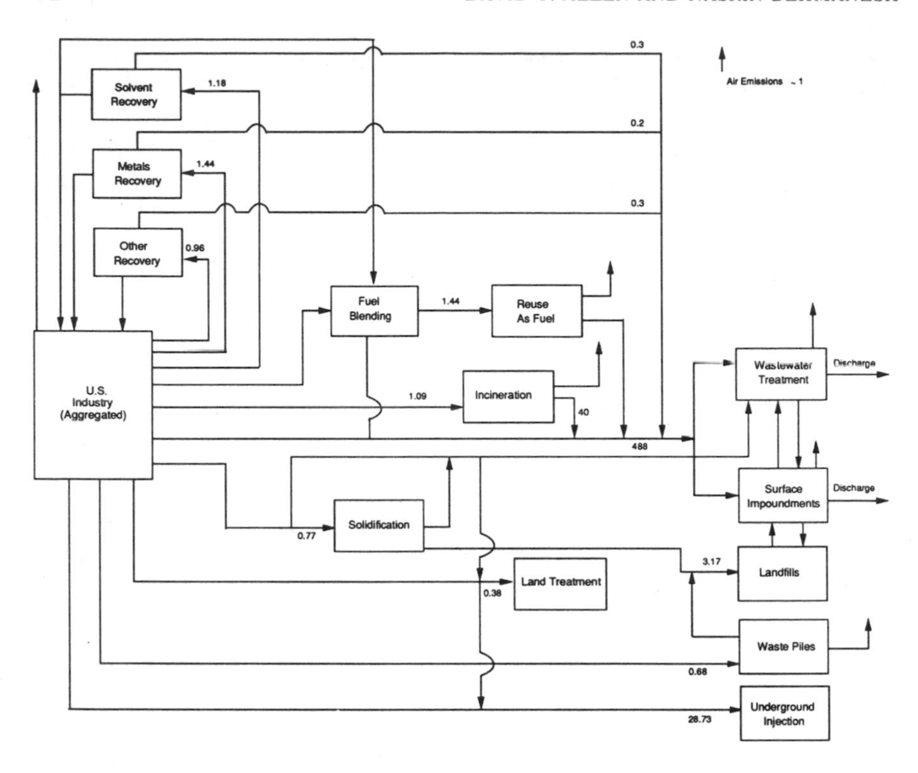

FIGURE 2 Flow of industrial hazardous wastes in treatment operations (1986 data in millions of tons per year).

components being recycled. A third set of waste streams, about 4 mt/yr, is sent to various thermal treatment technologies which include direct incineration, fuel blending, and reuse as fuel.

While an examination of total waste flows is a necessary first step in assessing the use of waste streams as raw materials, total mass is not a good indicator of the potential value of waste streams. Instead, the concentration and mass flow rates of valuable resources in the waste streams will be the most important evaluation criteria. Unfortunately, reliable composition data are not generally available for waste streams, so it is not always possible to evaluate the potential for recycling. One set of materials for which waste composition data are available is metals. The National Hazardous Waste Survey contains data on the flow rates of the 16 metals listed in CFR 261, Appendix VIII (Resource Conservation and Recovery Act, Subtitle C, Sections 3001–3013, 42 U.S.C., Sections 6921–6934 [1976] and Supplement IV [1980]). The next two sections will review the mass flow rates and concentration distributions of selected metals in industrial waste streams.

FLOWS OF SELECTED METALS IN INDUSTRIAL WASTES

As a first step in evaluating the potential of industrial wastes for use as raw materials, we will consider the flows of three metals—cadmium, chromium, and lead. Figures 3, 4, and 5 report the amount of cadmium, chromium, and lead sent to major industrial waste management operations. For cadmium and chromium, only a small fraction of the material is recovered. In the case of cadmium, approximately 1,300 out of a possible 16,000 tons were sent to recovery operations in 1986. In contrast, a major portion of lead generated as industrial wastes (106,000 out of 189,000 tons) is sent to metal recovery. Recycling is feasible for many lead-containing streams because an efficient collection and reprocessing system is in place for used automotive storage batteries. With such extensive recycling of lead, it should come as no surprise that secondary nonferrous metal processing (Standard Industrial Classification [SIC] code 3341), which is largely lead battery recycling, is the dominant source of lead wastes (Figure 6).

The flows of cadmium, chromium, and lead in industrial hazardous wastes,

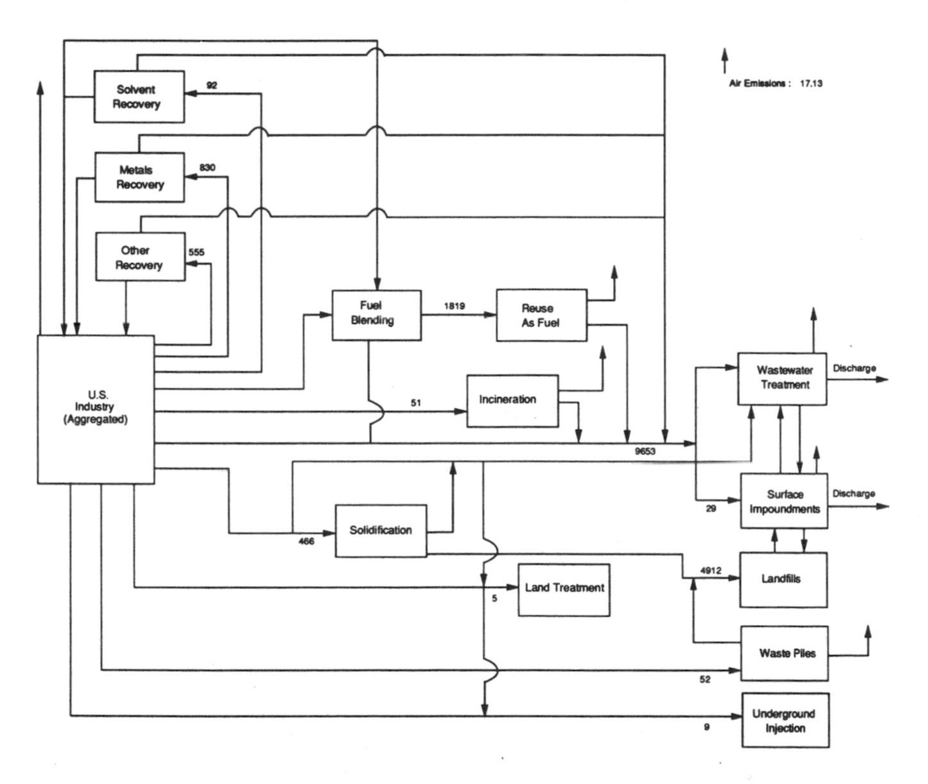

FIGURE 3 Flow of cadmium in hazardous waste treatment operations in 1986 (tons per year).

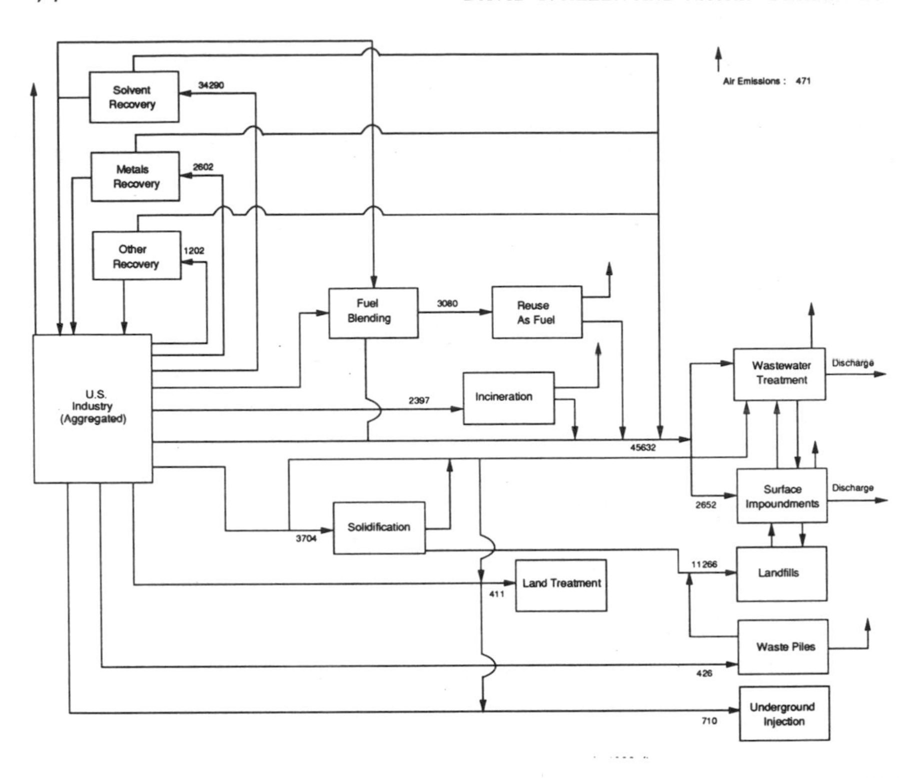

FIGURE 4 Flow of chromium in hazardous waste treatment operations in 1986 (tons per year).

illustrated in Figures 2–5, cannot be considered in isolation. They are merely a part, albeit a major part, of the total waste stream flow. A second major component of waste stream flows is municipal solid waste. The total mass flow of municipal solid waste is approximately 0.2 billion tons per year, which is considerably less than the total flow of industrial hazardous wastes (0.75 billion tons per year) and total industrial wastes (12 billion tons per year). Simple tonnage comparisons can be misleading, however. To assess the potential value of industrial hazardous waste and municipal solid wastes as sources of raw materials, it is necessary to compare the flows of specific materials. For example, according to the U.S. EPA (1989), the dominant contributor to lead in the municipal solid waste (MSW) stream is storage batteries. Depending on rates of recycling, batteries contribute between 100,000 and 150,000 tons per year of lead to MSW. Recent data (EPA, 1989) indicate a 90 percent rate of recycling, resulting in 100,000 tons of waste out of the 800,000 tons of lead in used batteries. Other sources of lead in the MSW stream include consumer electronics (estimated to be 60,000 tons), glass and ceramics (8,000 tons), plastics (4,000 tons), metals such as soldered cans

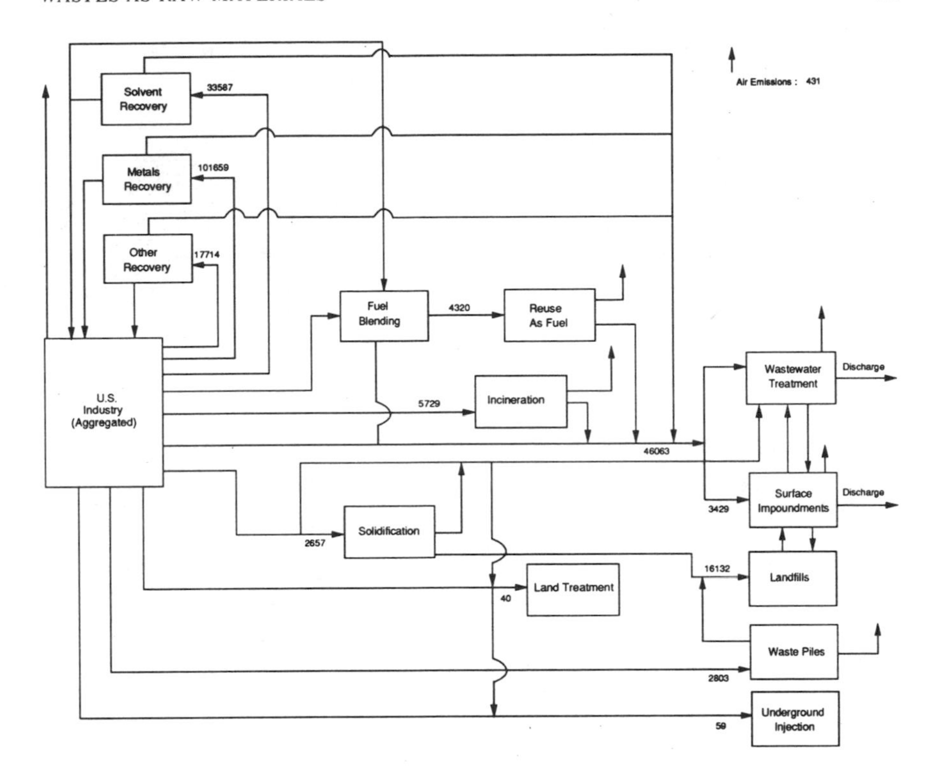

FIGURE 5 Flow of lead in hazardous waste treatment operations in 1986 (tons per year).

(1,000 tons) and pigments (1,000 tons). None of these other sources of waste result in any significant degree of lead recycling, largely because of the low concentrations of lead in the products.

Analysis of these waste management data represents a first step in performing studies in industrial ecology. The next step is to integrate waste generation data with production data. Coupling the waste flow data for lead presented above with the data on lead production and consumption presented in Table 1 yields the lead flow diagram shown in Figure 7. Lead is used at a rate of roughly 1.2 million tons per year. Most is consumed in the production of lead storage batteries and these batteries are eventually retired. Roughly 90 percent of these used batteries are recycled, so the net loss of lead through battery disposal is about 100,000 tons. Other sources of lead in MSW total roughly 70,000 tons per year. Industrial hazardous wastes are another sink for lead. Of the 189,000 tons of lead in hazardous wastes, roughly 53,000 tons are sent to disposal. The remainder is recycled, but the recycling of both hazardous and battery wastes generates roughly 100,000 tons of lead waste. Comparing the total flow of lead with the amount of lead

CADMIUM

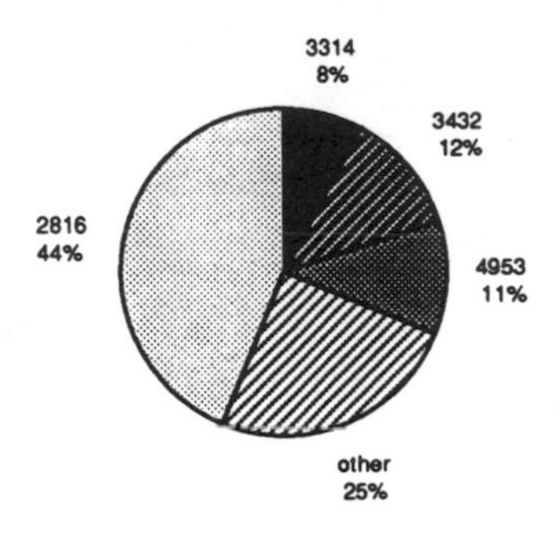

SIC	INDUSTRY
2816:	Inorganic pigments
3314:	Primary metals industries
3432:	Plumbing, fitting, & brass goods
4953:	Refuse systems

Total: 16,100 tons/year

CHROMIUM

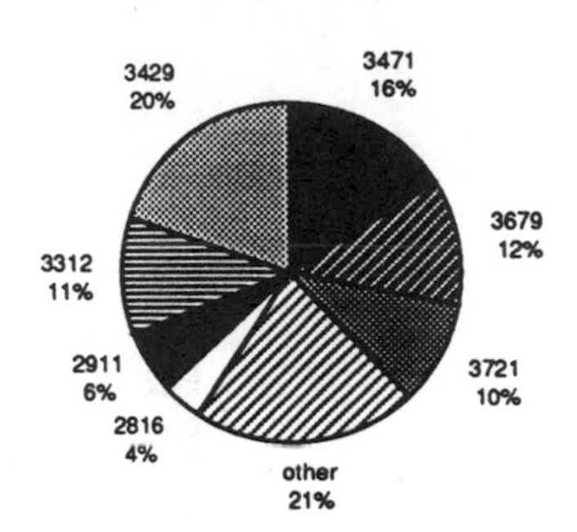

SIC	INDUSTRY
2816:	Inorganic pigments
2911:	Petroleum refining
3312:	Blast furnace & steel mills
3429:	Hardware, nec
3471:	Plating & polishing
3679:	Electronic components, nec
3721:	Aircraft

Total: 89,900 tons/year

LEAD

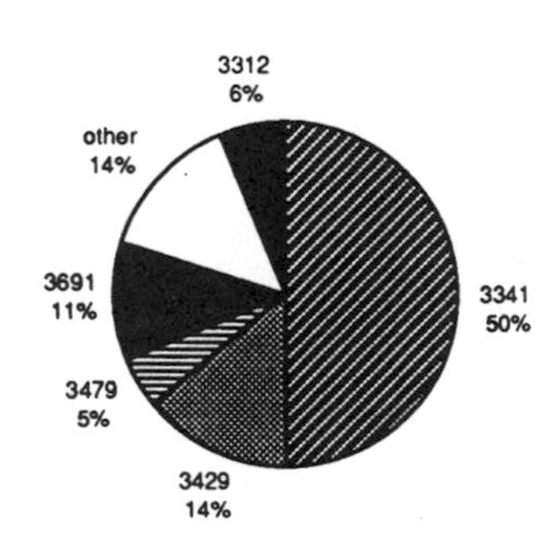

SIC	INDUSTRY
3312:	Blast furnace & steel mills
3341:	Secondary nonferrous metals
3429:	Hardware, nec
3479:	Metal coating & allied services
3691:	Storage batteries

Total: 188,600 tons/year

FIGURE 6 Industrial sources of cadmium, chromium, and lead wastes.

TABLE 1 Production and Consumption of Lead, 1986 and 1989

	Amount, 1986[a] (metric tons)	Percent, 1986[a]	Amount, 1989[b] (metric tons)	Percent, 1989[b]
Production				
Mine production	340,000		411,000	
Secondary lead	615,000		809,000	
Consumption				
Metal products	146,000	12.7	170,000	13.3
Storage batteries	854,000	77.9	1,012,000	78.9
Other oxides	69,000	5.0	58,000	4.5
Miscellaneous (including gasoline additives)	55,000	4.4	43,000	3.3
TOTAL	1,124,000	100.0	1,283,000	100.00

[a]U.S. Department of the Interior (1988).
[b]U.S. Department of the Interior (1991).

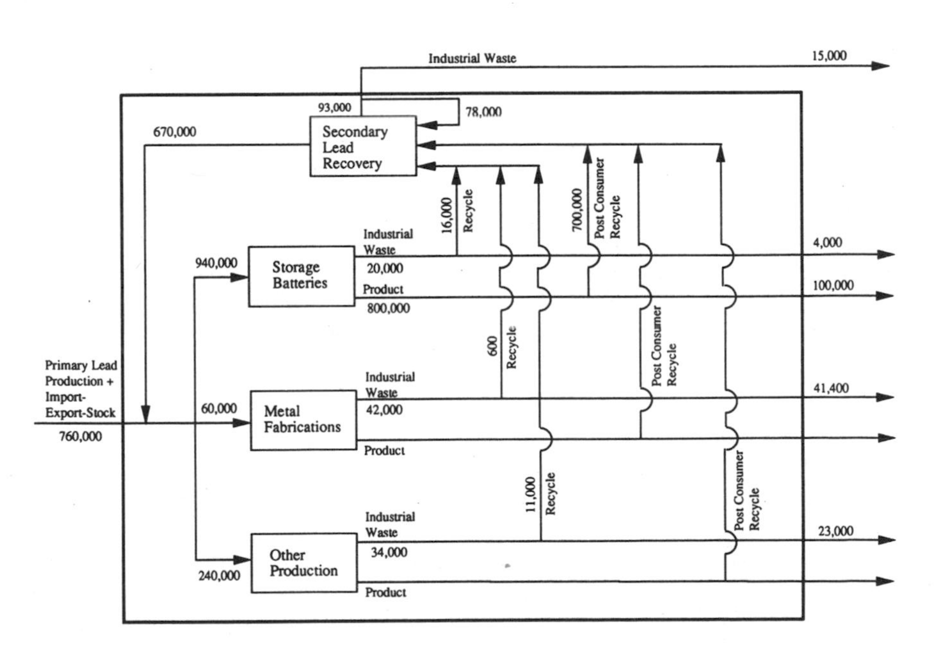

FIGURE 7 Simplified model of the industrial ecology of lead (amounts in tons per year).

eventually requiring disposal, we see that lead is reprocessed at about 75 percent efficiency:

$$\% \text{ Efficiency} = 100 \times 1 - (300{,}000 \text{ tons discarded} \div 1{,}200{,}000 \text{ tons consumed}) = 75\%$$

If this efficiency is to be improved, then the streams that are currently reaching disposal must find productive use. Overall recycling efficiency could be improved by increasing the collection of lead batteries above 90 percent, by improving the efficiency of secondary lead smelting, and by targeting for recycling industrial waste streams from nonbattery operations. Figure 8 compares some of the waste streams currently requiring disposal with those currently being recycled. Examination of Figure 8 reveals that more concentrated waste streams are more likely to be recycled than waste streams with low lead concentration. Although the decision whether or not to reclaim a metal from a waste stream is complex, it is in essence an economic question. Accordingly, it depends not only on the value of the recycled material, but to a significant extent on concentration.

DILUTION DETERMINES RECYCLABILITY

As in the case of lead, the concentration at which recycling of other materials in the waste stream becomes cost-effective depends on the value of the raw material (see Figure 9). The Sherwood diagram (Figure 1) showed that whereas materials such as gold and radium can be recovered from raw materials that are quite dilute in the resource, materials such as copper can be recovered economically only from relatively rich ores.

Given the price, it is therefore possible to estimate the concentration at which materials can be recovered. By comparing metal prices, minimum economically recoverable concentration (from the Sherwood diagram), and data on the concentration distributions of metals in waste streams (Figure 9), it is possible to estimate what fraction of metals in hazardous waste streams can be recycled. These estimates are reported in Table 2 and indicate that metals in hazardous wastes are underutilized. This could be because only waste streams with very high metal concentrations are recovered or because only a small fraction of potential recyclers at all feasible concentration levels recover metals. Figure 10 is an attempt to differentiate between these two cases.

To develop Figure 10, the concentration distribution of metals in recycled waste streams was examined. The concentration below which only 10 percent of the metal recycling took place was assumed to be a lower bound for economic metal recovery from the waste. This concentration was then plotted, together with the 1986 metal price (recall that the waste data are from 1986) to generate Figure 10. Also plotted on Figure 10 is the Sherwood diagram for virgin materials. Comparison of the Sherwood plot for virgin materials and the waste concentration data

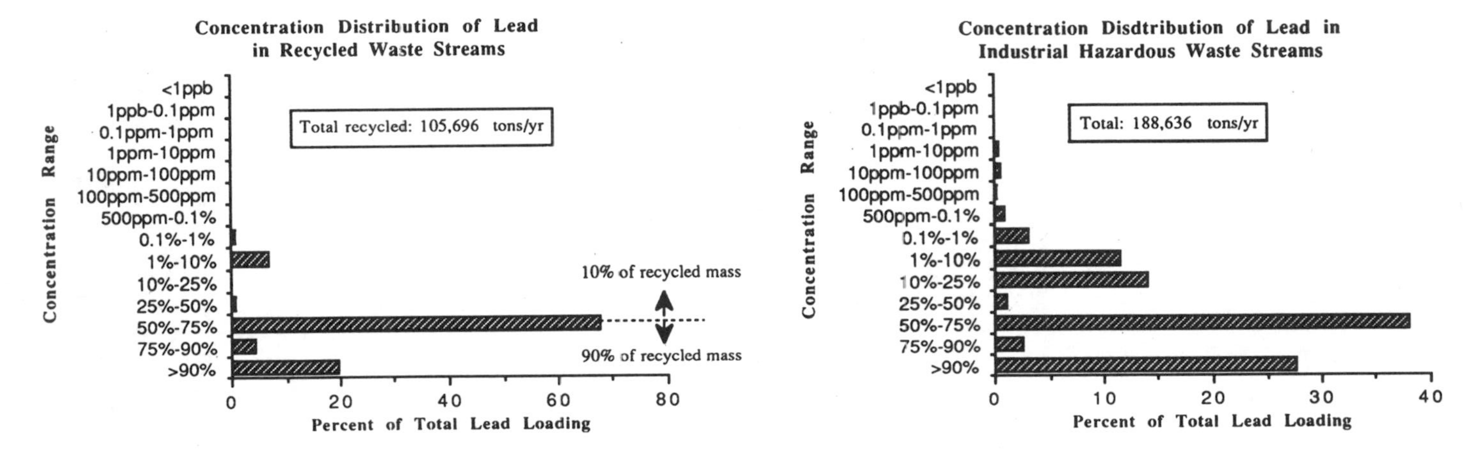

FIGURE 8 Concentration distributions of lead in waste streams undergoing recycling and concentration distributions of lead in all industrial hazardous waste streams (1986). The concentration below which only 10 percent of recyling takes place is noted.

TABLE 2 Percentage of Total Metal Loadings That Can Be Recovered Economically (derived from Sherwood Plot) from Industrial Hazardous Waste Streams

Metal	Minimum Concentration Recoverable, from Sherwood Plot (mass fraction)	Percent of Metal Theoretically Recoverable (%)	Percent Recycled in 1986 (%)
Sb	0.00405	74-87	32
As	0.00015	98-99	3
Ba	0.0015	95-98	4
Be	0.012	54-84	31
Cd	0.0048	82-97	7
Cr	0.0012	68-89	8
Cu	0.0022	85-92	10
Pb	0.074	84-95	56
Hg	0.00012	99	41
Ni	0.0066	100	0.1
Se	0.0002	93-95	16
Ag	0.000035	99-100	1
Tl	0.00004	97-99	1
V	0.0002	74-98	1
Zn	0.0012	96-98	13

reveals that most metals in waste streams are recycled only at high concentrations. The concentration of resources in recycled wastes is generally higher than for virgin materials, indicating significant disincentives to make use of waste. Figures 9 and 10 demonstrate that there are many opportunities for increased recycling.

SUMMARY

Compositions and sources of recycled waste streams can be examined and opportunities for improving recycling efficiencies can be explored. Unfortunately, these analyses rely extensively on a single data base of industrial waste, the National Hazardous Waste Survey. This collection of data, which is the only comprehensive and detailed source of information on the composition of industrial hazardous waste streams, was based on wastes generated during 1986 and is already somewhat outdated. The data are also restricted to wastes classified as hazardous under the provisions of the Resource Conservation and Recovery Act. The lack of current, comprehensive, and reliable data on waste composition remains a serious barrier to studies in industrial ecology.

If the limitations of available data are understood, however, it is possible to

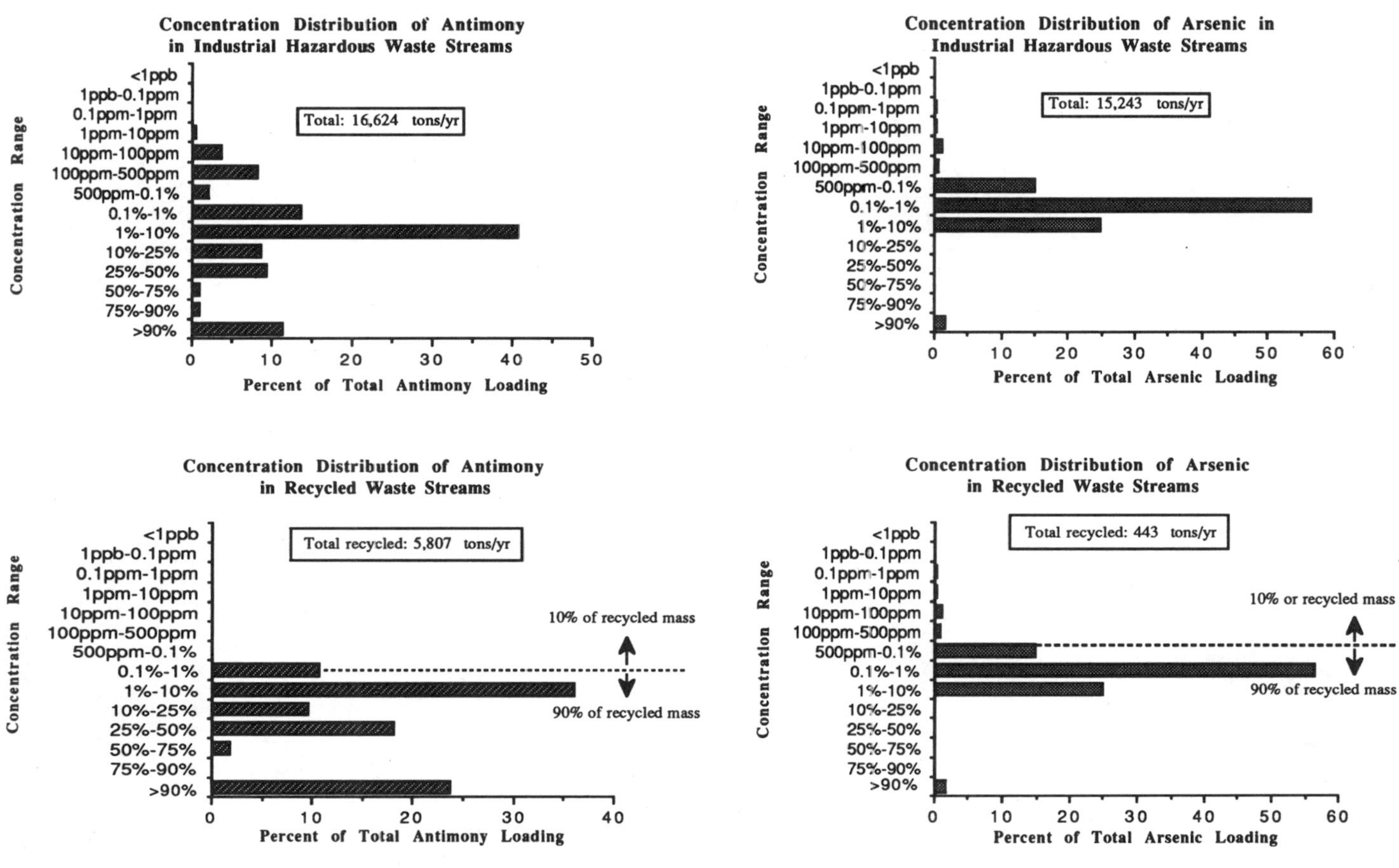

FIGURE 9 Concentration distribution of recycled metals and the concentration distributions of metals in all industrial hazardous waste streams (1986 data). The concentration below which only 10 percent of recycling takes place is noted for each metal.

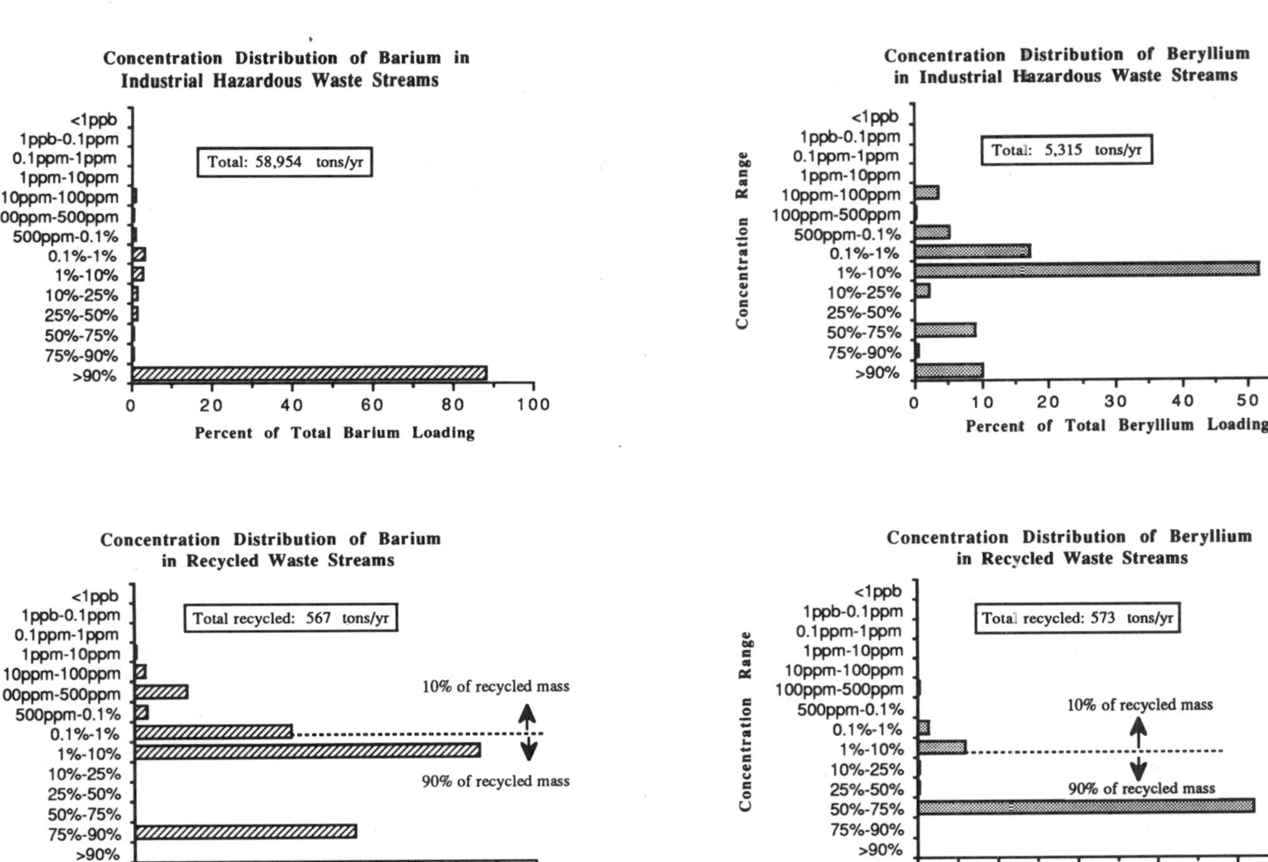
Concentration Distribution of Barium in Industrial Hazardous Waste Streams
Total: 58,954 tons/yr
Concentration Range
<1ppb
1ppb-0.1ppm
0.1ppm-1ppm
1ppm-10ppm
10ppm-100ppm
100ppm-500ppm
500ppm-0.1%
0.1%-1%
1%-10%
10%-25%
25%-50%
50%-75%
75%-90%
>90%
0
20
40
60
80
100
Percent of Total Barium Loading
Concentration Distribution of Beryllium in Industrial Hazardous Waste Streams
Total: 5,315 tons/yr
Concentration Range
<1ppb
1ppb-0.1ppm
0.1ppm-1ppm
1ppm-10ppm
10ppm-100ppm
100ppm-500ppm
500ppm-0.1%
0.1%-1%
1%-10%
10%-25%
25%-50%
50%-75%
75%-90%
>90%
0
10
20
30
40
50
60
Percent of Total Beryllium Loading
Concentration Distribution of Barium in Recycled Waste Streams
Total recycled: 567 tons/yr
10% of recycled mass
90% of recycled mass
Concentration Range
<1ppb
1ppb-0.1ppm
0.1ppm-1ppm
1ppm-10ppm
10ppm-100ppm
100ppm-500ppm
500ppm-0.1%
0.1%-1%
1%-10%
10%-25%
25%-50%
50%-75%
75%-90%
>90%
0
10
20
30
40
50
Percent of Total Barium Loading
Concentration Distribution of Beryllium in Recycled Waste Streams
Total recycled: 573 tons/yr
10% of recycled mass
90% of recycled mass
Concentration Range
<1ppb
1ppb-0.1ppm
0.1ppm-1ppm
1ppm-10ppm
10ppm-100ppm
100ppm-500ppm
500ppm-0.1%
0.1%-1%
1%-10%
10%-25%
25%-50%
50%-75%
75%-90%
>90%
0
20
40
60
80
100
Percent of Total Beryllium Loading

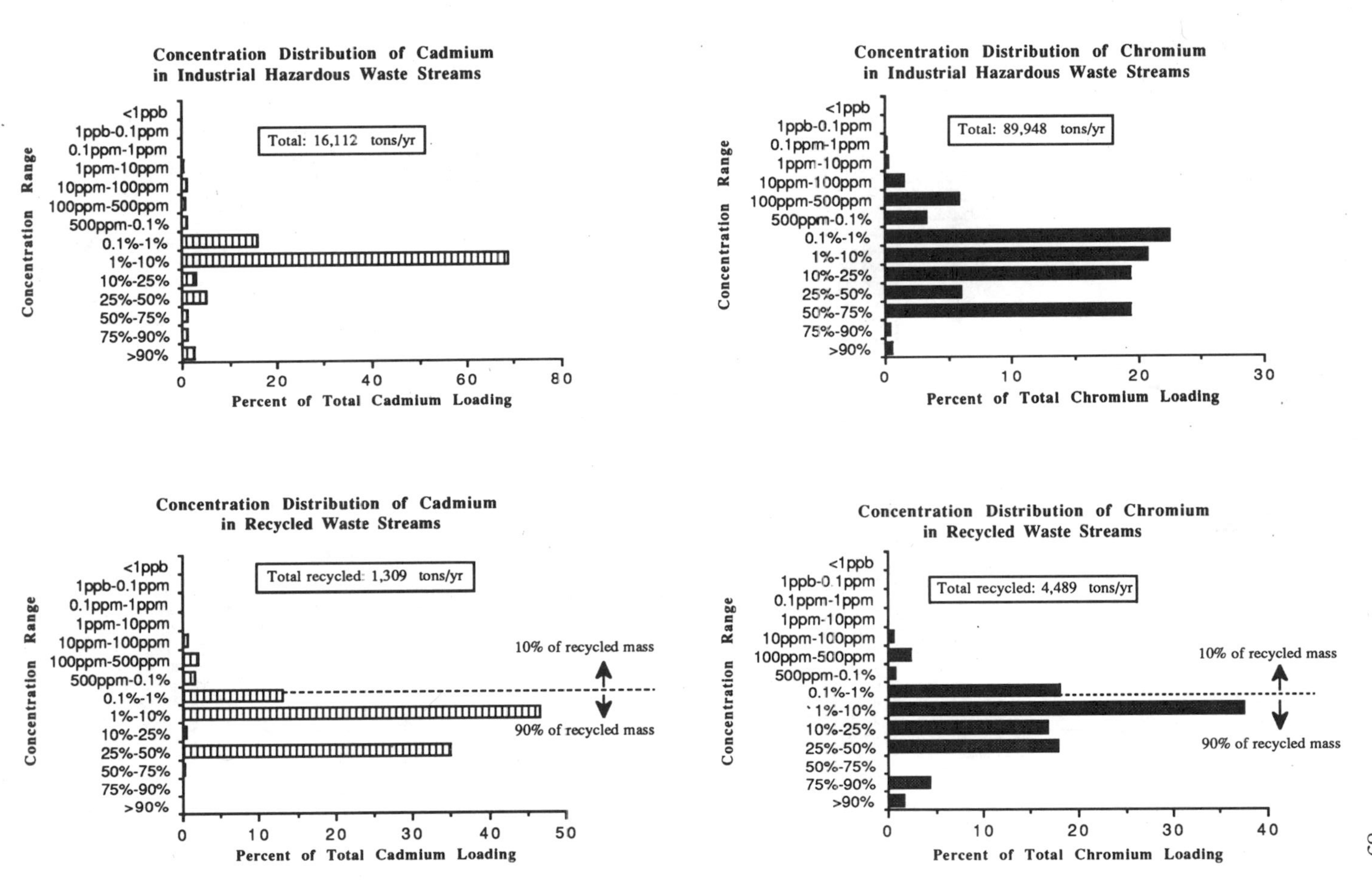
Concentration Distribution of Cadmium in Industrial Hazardous Waste Streams
Total: 16,112 tons/yr
Concentration Range
<1ppb
1ppb-0.1ppm
0.1ppm-1ppm
1ppm-10ppm
10ppm-100ppm
100ppm-500ppm
500ppm-0.1%
0.1%-1%
1%-10%
10%-25%
25%-50%
50%-75%
75%-90%
>90%
0
20
40
60
80
Percent of Total Cadmium Loading
Concentration Distribution of Chromium in Industrial Hazardous Waste Streams
Total: 89,948 tons/yr
Concentration Range
<1ppb
1ppb-0.1ppm
0.1ppm-1ppm
1ppm-10ppm
10ppm-100ppm
100ppm-500ppm
500ppm-0.1%
0.1%-1%
1%-10%
10%-25%
25%-50%
50%-75%
75%-90%
>90%
0
10
20
30
Percent of Total Chromium Loading
Concentration Distribution of Cadmium in Recycled Waste Streams
Total recycled: 1,309 tons/yr
10% of recycled mass
90% of recycled mass
Concentration Range
<1ppb
1ppb-0.1ppm
0.1ppm-1ppm
1ppm-10ppm
10ppm-100ppm
100ppm-500ppm
500ppm-0.1%
0.1%-1%
1%-10%
10%-25%
25%-50%
50%-75%
75%-90%
>90%
0
10
20
30
40
50
Percent of Total Cadmium Loading
Concentration Distribution of Chromium in Recycled Waste Streams
Total recycled: 4,489 tons/yr
10% of recycled mass
90% of recycled mass
Concentration Range
<1ppb
1ppb-0.1ppm
0.1ppm-1ppm
1ppm-10ppm
10ppm-100ppm
100ppm-500ppm
500ppm-0.1%
0.1%-1%
1%-10%
10%-25%
25%-50%
50%-75%
75%-90%
>90%
0
10
20
30
40
Percent of Total Chromium Loading

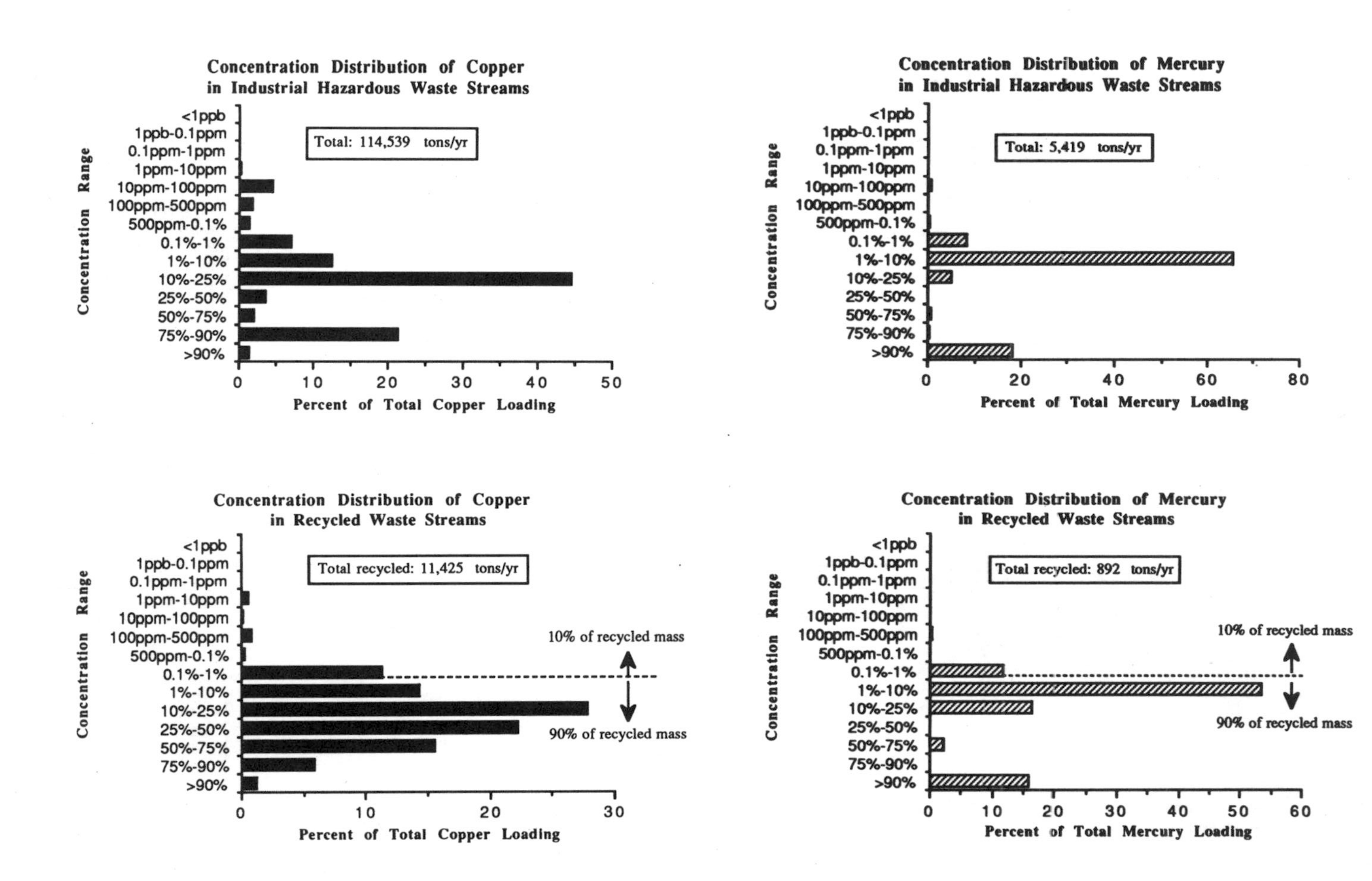
Concentration Distribution of Copper in Industrial Hazardous Waste Streams
Total: 114,539 tons/yr
Concentration Range
<1ppb
1ppb-0.1ppm
0.1ppm-1ppm
1ppm-10ppm
10ppm-100ppm
100ppm-500ppm
500ppm-0.1%
0.1%-1%
1%-10%
10%-25%
25%-50%
50%-75%
75%-90%
>90%
0 10 20 30 40 50
Percent of Total Copper Loading
Concentration Distribution of Mercury in Industrial Hazardous Waste Streams
Total: 5,419 tons/yr
0 20 40 60 80
Percent of Total Mercury Loading
Concentration Distribution of Copper in Recycled Waste Streams
Total recycled: 11,425 tons/yr
10% of recycled mass
90% of recycled mass
0 10 20 30
Percent of Total Copper Loading
Concentration Distribution of Mercury in Recycled Waste Streams
Total recycled: 892 tons/yr
10% of recycled mass
90% of recycled mass
0 10 20 30 40 50 60
Percent of Total Mercury Loading

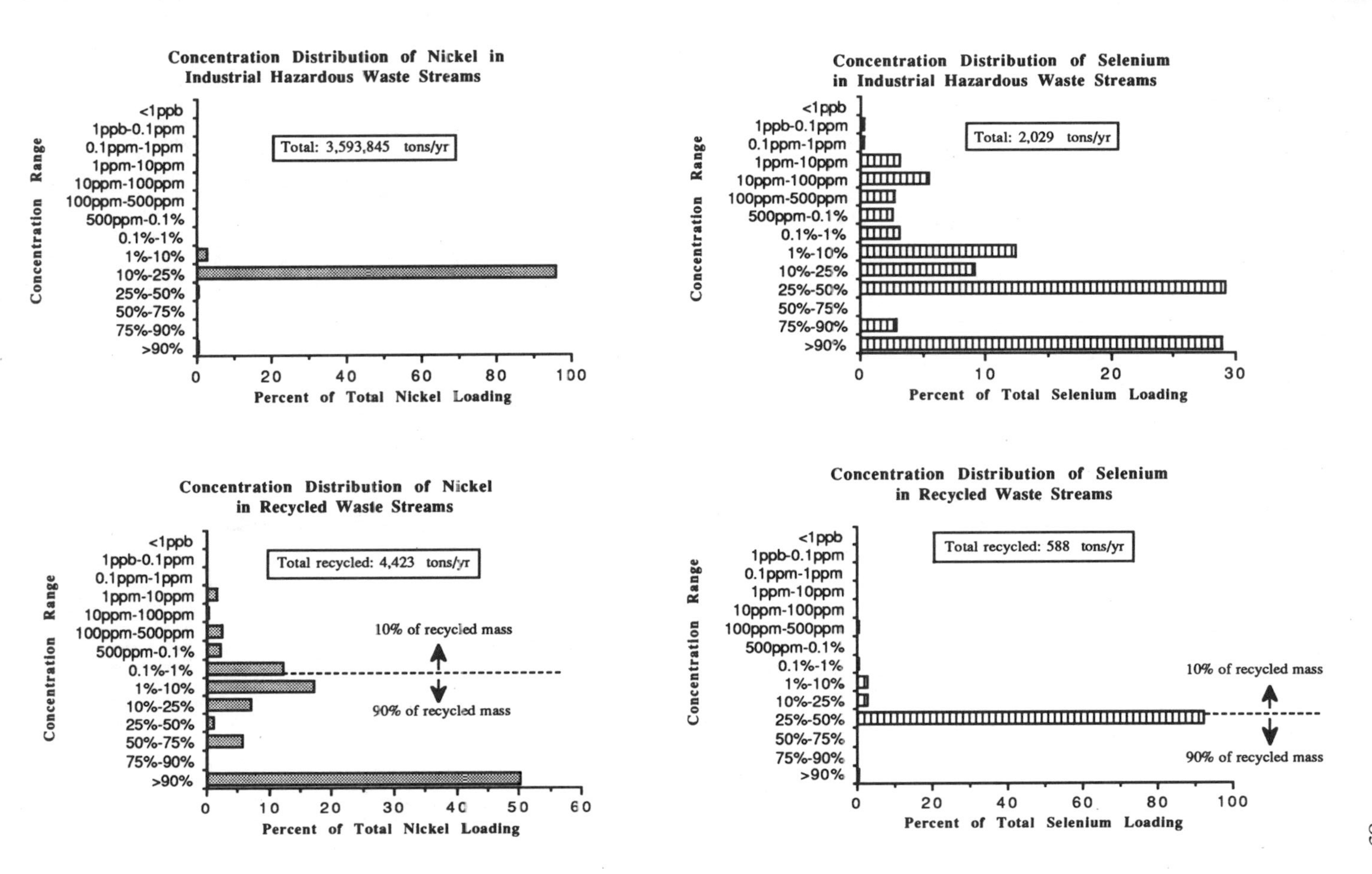
Concentration Distribution of Nickel in Industrial Hazardous Waste Streams
Total: 3,593,845 tons/yr
Concentration Range
<1ppb
1ppb-0.1ppm
0.1ppm-1ppm
1ppm-10ppm
10ppm-100ppm
100ppm-500ppm
500ppm-0.1%
0.1%-1%
1%-10%
10%-25%
25%-50%
50%-75%
75%-90%
>90%
0
20
40
60
80
100
Percent of Total Nickel Loading
Concentration Distribution of Selenium in Industrial Hazardous Waste Streams
Total: 2,029 tons/yr
Concentration Range
0
10
20
30
Percent of Total Selenium Loading
Concentration Distribution of Nickel in Recycled Waste Streams
Total recycled: 4,423 tons/yr
Concentration Range
10% of recycled mass
90% of recycled mass
0
10
20
30
40
50
60
Percent of Total Nickel Loading
Concentration Distribution of Selenium in Recycled Waste Streams
Total recycled: 588 tons/yr
Concentration Range
10% of recycled mass
90% of recycled mass
0
20
40
60
80
100
Percent of Total Selenium Loading

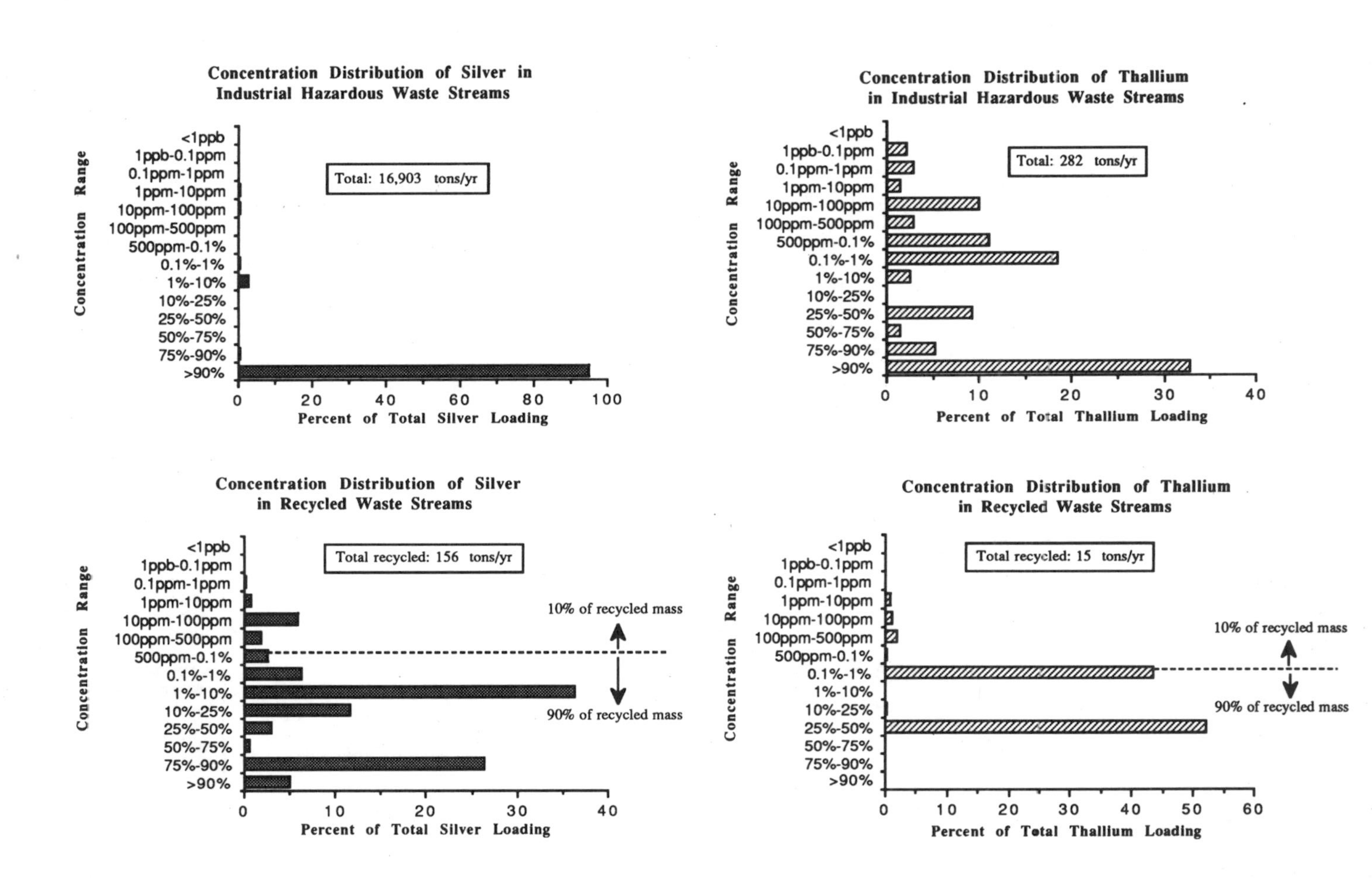
Concentration Distribution of Silver in Industrial Hazardous Waste Streams
Total: 16,903 tons/yr
Concentration Range
<1ppb
1ppb-0.1ppm
0.1ppm-1ppm
1ppm-10ppm
10ppm-100ppm
100ppm-500ppm
500ppm-0.1%
0.1%-1%
1%-10%
10%-25%
25%-50%
50%-75%
75%-90%
>90%
0
20
40
60
80
100
Percent of Total Silver Loading
Concentration Distribution of Thallium in Industrial Hazardous Waste Streams
Total: 282 tons/yr
Concentration Range
<1ppb
1ppb-0.1ppm
0.1ppm-1ppm
1ppm-10ppm
10ppm-100ppm
100ppm-500ppm
500ppm-0.1%
0.1%-1%
1%-10%
10%-25%
25%-50%
50%-75%
75%-90%
>90%
0
10
20
30
40
Percent of Total Thallium Loading
Concentration Distribution of Silver in Recycled Waste Streams
Total recycled: 156 tons/yr
10% of recycled mass
90% of recycled mass
Concentration Range
<1ppb
1ppb-0.1ppm
0.1ppm-1ppm
1ppm-10ppm
10ppm-100ppm
100ppm-500ppm
500ppm-0.1%
0.1%-1%
1%-10%
10%-25%
25%-50%
50%-75%
75%-90%
>90%
0
10
20
30
40
Percent of Total Silver Loading
Concentration Distribution of Thallium in Recycled Waste Streams
Total recycled: 15 tons/yr
10% of recycled mass
90% of recycled mass
Concentration Range
<1ppb
1ppb-0.1ppm
0.1ppm-1ppm
1ppm-10ppm
10ppm-100ppm
100ppm-500ppm
500ppm-0.1%
0.1%-1%
1%-10%
10%-25%
25%-50%
50%-75%
75%-90%
>90%
0
10
20
30
40
50
60
Percent of Total Thallium Loading

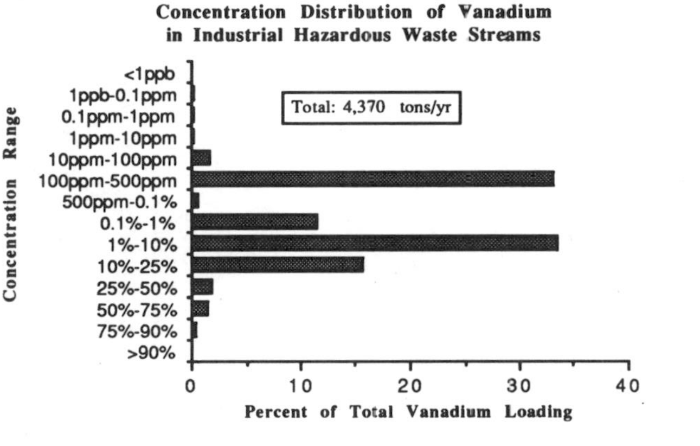
Concentration Distribution of Vanadium in Industrial Hazardous Waste Streams
Total: 4,370 tons/yr
Concentration Range
<1ppb
1ppb-0.1ppm
0.1ppm-1ppm
1ppm-10ppm
10ppm-100ppm
100ppm-500ppm
500ppm-0.1%
0.1%-1%
1%-10%
10%-25%
25%-50%
50%-75%
75%-90%
>90%
0
10
20
30
40
Percent of Total Vanadium Loading

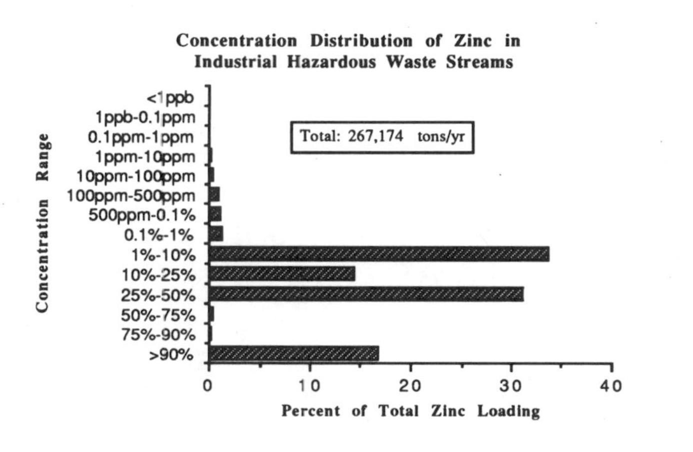
Concentration Distribution of Zinc in Industrial Hazardous Waste Streams
Total: 267,174 tons/yr
Concentration Range
<1ppb
1ppb-0.1ppm
0.1ppm-1ppm
1ppm-10ppm
10ppm-100ppm
100ppm-500ppm
500ppm-0.1%
0.1%-1%
1%-10%
10%-25%
25%-50%
50%-75%
75%-90%
>90%
0
10
20
30
40
Percent of Total Zinc Loading

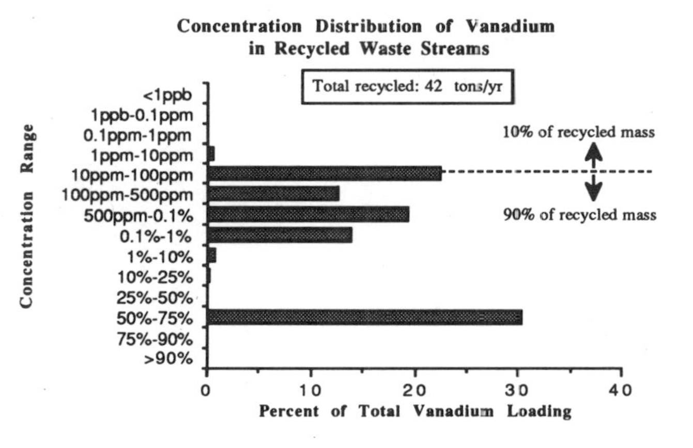
Concentration Distribution of Vanadium in Recycled Waste Streams
Total recycled: 42 tons/yr
10% of recycled mass
90% of recycled mass
Concentration Range
<1ppb
1ppb-0.1ppm
0.1ppm-1ppm
1ppm-10ppm
10ppm-100ppm
100ppm-500ppm
500ppm-0.1%
0.1%-1%
1%-10%
10%-25%
25%-50%
50%-75%
75%-90%
>90%
0
10
20
30
40
Percent of Total Vanadium Loading

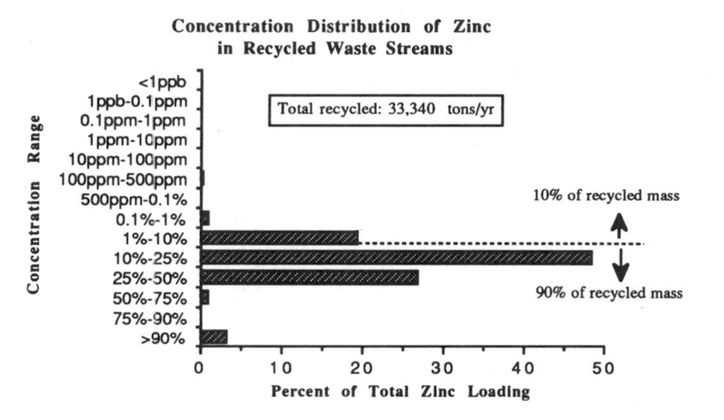
Concentration Distribution of Zinc in Recycled Waste Streams
Total recycled: 33,340 tons/yr
10% of recycled mass
90% of recycled mass
Concentration Range
<1ppb
1ppb-0.1ppm
0.1ppm-1ppm
1ppm-10ppm
10ppm-100ppm
100ppm-500ppm
500ppm-0.1%
0.1%-1%
1%-10%
10%-25%
25%-50%
50%-75%
75%-90%
>90%
0
10
20
30
40
50
Percent of Total Zinc Loading

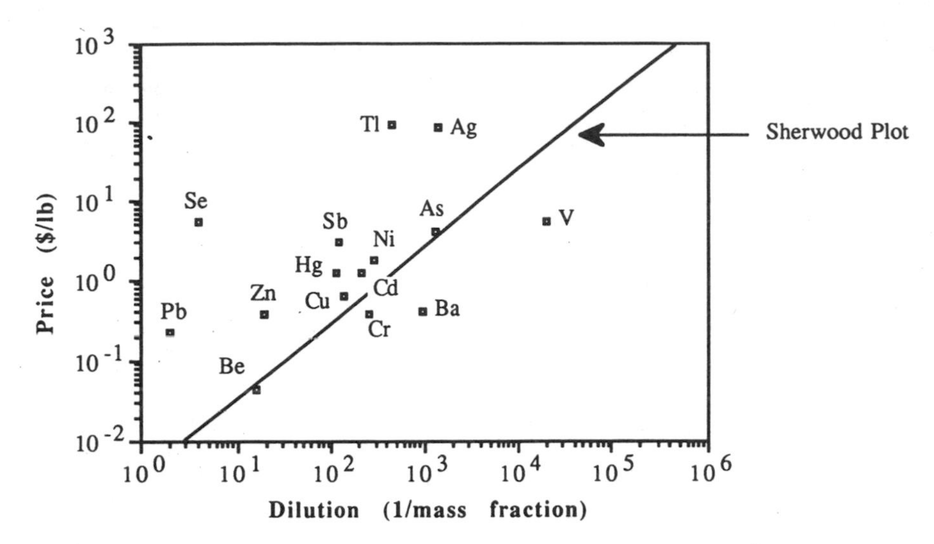

FIGURE 10 The Sherwood plot for waste streams. The minimum concentration of metal wastes undergoing recycling (see Figures 8 and 9) is plotted against metal price. The Sherwood plot for virgin materials is provided by comparison. Points lying above the Sherwood plot indicate that the metals in the waste streams are underused, that is, waste streams undergoing disposal are richer than typical virgin materials. Points lying below the Sherwood plot indicate that the waste streams are vigorously recycled.

examine the industrial ecology of some metals. The results reveal that the concentrations of metal resources in many waste streams that are currently undergoing disposal are higher than for typical virgin resources. Thus, extensive waste trading could significantly reduce the quantity of waste requiring disposal.

ACKNOWLEDGMENT

This work was supported by the University of California Toxic Substances Research and Teaching Program.

REFERENCES

Allen, D. T., and R. Jain, eds. 1992. Special issue on industrial waste generation and management. Hazardous Waste and Hazardous Materials 9(1):1–111.

Eisenhauer, J., and Cordes, R. 1992. Industrial waste databases: A simple roadmap. Hazardous Waste and Hazardous Materials 9(1):1–19.

National Research Council. 1987. Separation and Purification: Critical Needs and Opportunities. Washington, D.C.: National Academy Press.

U.S. Department of the Interior. 1988. Minerals Yearbook: 1986, Metals and Minerals, Volume 1. Washington, D.C.: U.S. Government Printing Office.

U.S. Department of the Interior. 1991. Minerals Yearbook: 1989, Metals and Minerals, Volume 1. Washington, D.C.: U.S. Government Printing Office.

U.S. Environmental Protection Agency. 1988a. Report to Congress: Solid Waste Disposal in the United States, Volume 1, EPA 530-SW-88-011.

U.S. Environmental Protection Agency. 1988b. Report to Congress: Solid Waste Disposal in the United States, Volume 2, EPA 530-SW-88-011B.

U.S. Environmental Protection Agency. 1988c. 1986 National Survey of Hazardous Waste Treatment, Storage, Disposal and Recycling Facilities, EPA/530-SW-88-035.

U.S. Environmental Protection Agency. 1989. Characterization of Products Containing Lead and Cadmium in Municipal Solid Waste in the United States, 1970 to 2000, EPA/530-SW-89-015A.

U.S. Environmental Protection Agency. 1990. Characterization of Municipal Solid Waste in the United States: 1990 Update, EPA 530-SW-90-042.

U.S. Environmental Protection Agency. 1991. Toxics in the Community, EPA 560/4-91/014.

The Greening of Industrial Ecosystems. 1994.
Pp. 90–97. Washington, DC:
National Academy Press.

Economics and Sustainable Development

PIERRE CROSSON and MICHAEL A. TOMAN

Sustainable development has become a new watchword for assessing human impacts on the natural environment and resource base. A concern that economic development, exploitation of natural resources, and infringement on environmental resources are not sustainable is expressed more and more frequently in analytical studies, conferences, and policy debates. To identify what may be required to achieve sustainability, however, it is necessary to have a clear understanding of what the concept means. In particular, it is necessary to understand what is new in the concept of sustainable development as distinct from concepts that already are well established in environmental and natural resource economics.

The World Commission on Environment and Development (known popularly as the Brundtland Commission) labeled sustainable development in its 1987 report *Our Common Future* as "development that meets the needs of the present without compromising the ability of future generations to meet their own needs." Thus, sustainability involves concern for the interests of future generations.

A second common feature in discussion of sustainability is the question of substitutability between natural resources (including the environment) and human-made capital, including human capital itself. The Brundtland report foresees "the possibility for a new era of economic growth, one that must be based on policies that sustain and expand the environmental resource base." This view gives special emphasis to the natural endowment but sees preservation of this endowment as entirely consistent with economic growth. Some scholars, notably the economist Julian Simon (1981), even question whether sustainability is a significant issue, pointing out that humankind consistently has managed in the past to avoid the specter of Malthusian scarcity through resource substitution and technical ingenuity.

Others, however, notably the ecologists Paul and Anne Ehrlich (1990) and the economist Herman Daly (see Daly and Cobb, 1989), believe that the scale of human pressure on natural systems already is well past a sustainable level. They point out that the world's human population most likely will at least double before stabilizing, and that to achieve any semblance of a decent living standard for the majority of people, the current level of world economic activity must grow, perhaps fivefold to tenfold. They cannot conceive of already stressed ecological systems tolerating the intense flows of materials use and waste discharge that presumably would be required to accomplish this growth, no matter what level of investment in built capital and technology occurs. On the contrary, an implication of the Ehrlich/Daly argument is that even the present global population and economy are unsustainable.

KEY CONCEPTUAL ISSUES

As noted above, intergenerational fairness is a key component of sustainability. The standard approach to intergenerational trade-offs in economics involves assigning benefits and costs according to some representative set of individual preferences, and discounting costs and benefits accruing to future generations just as future receipts and burdens experienced by members of the current generation are discounted. The justifications for discounting over time are first, that people prefer current benefits over future benefits (and weigh current costs more heavily than future costs); and second, that receipts in the future are less valuable than current receipts from the standpoint of the current decision maker, because current receipts can be invested to increase capital and future income.

Critics of the standard approach take issue with both rationales for unfettered application of discounting in an intergenerational context. They maintain that invoking impatience to justify discounting entails the exercise of the current generation's influence over future generations in ways that are ethically questionable. The capital growth argument for intergenerational discounting also is suspect, critics argue, because in many cases the environmental resources at issue—for example, the capacity of the atmosphere to safely absorb greenhouse gases or the extent of biological diversity—are seen to be inherently limited in supply. These criticisms do not imply that discounting should be abolished (especially since this could increase current exploitation of natural and environmental capital), but they do suggest that discounting might best be applied in tandem with safeguards on the integrity of key resources such as ecological life-support systems.

Critics also question whether the preferences of an "average" member of the current generation should be the sole or even primary guide to intergenerational resource trade-offs, particularly if some current resource uses threaten the future well-being of the entire human species. (Adherents of "deep ecology" even taken issue with putting human values at the center of the debate, arguing instead that other elements of the global ecological system have equal moral claims to be sustained.) The first part of the question is ill-taken: if anyone speaks for the interests of future generations, it inevitably must be members of the present generation. The second part of the question has content and is inherently hard to answer because members of the current generation cannot be sure, beyond a few generalities, of what will constitute the future well-being of humans. People to the farthest generation will need "adequate" nutrition, health care, education, and housing. But apart from problems of defining "adequate," these components of welfare by no means constitute the whole. Adequately fed, housed, educated, and healthy individuals will value a wide range of other services provided, directly and indirectly, by the natural environment and resource base. Our generation can only dimly perceive many of these values, and many more must be completely out of our sight.

A second key component of sustainability involves the specification of what is to be sustained. If one accepts that there is some collective responsibility of stewardship owed to future generations, what kind of "social capital" needs to be intergenerationally transferred to meet that obligation? One view, to which many economists are inclined, is that all resources—the natural endowment, physical capital, human knowledge, and abilities—are relatively fungible sources of well-being. Thus large-scale damages to ecosystems through drainage of wetlands, loss of species diversity, widespread deforestation, global warming, and so on are not intrinsically unacceptable from this point of view; the question is whether compensatory investments in other forms of capital are possible and are undertaken to protect the welfare of future generations. Investments in human knowledge, technique, and social organization are especially pertinent in evaluating these issues.

An alternative view, embraced by many ecologists and some economists, is that such compensatory investments often are infeasible as well as ethically indefensible. Physical laws are seen as limiting the extent to which other resources can be substituted for ecological degradation. Healthy ecosystems, including those that provide genetic diversity in relatively unmanaged environments, are seen as offering resilience against unexpected changes and preserving options for future generations. For these natural life-support systems, no practical substitutes are possible, in this view, and degradation may be irreversible. In such cases (and perhaps in others as well), compensation cannot be meaningfully specified. In addition, environmental quality may complement capital growth as a source of economic progress, particularly for poorer countries. Such complementarity argues against the notion of substituting human-made capital to compensate for natural degradation.

In considering resource substitutability, economists and ecologists often also differ on the appropriate geographical scale. Scale is important because opportunities for resource trade-offs generally are greater at the level of the nation or the globe than at the level of the individual community or regional ecosystem. However, a concern only with large regional or global aggregates may overlook unique attributes of particular local ecosystems or local constraints on resource substitution.

There also is sharp disagreement on the issue of the scale of human impact relative to global carrying capacity. As a crude caricature, it is generally true that economists are less inclined than ecologists to see this as a serious problem, putting more faith in the capacities of resource substitution (including substitution of knowledge for materials) and technical innovation to ameliorate scarcity. Rather than viewing it as an immutable constraint, economists regard carrying capacity as endogenous and dynamic. Ecologists, in contrast, emphasize that large-scale ecological impacts are precisely those that may be the most damaging and least irreversible.

THE SAFE MINIMUM STANDARD

Concerns over intergenerational fairness, resource constraints, and human scale provide a rationale for the concept of a "safe minimum standard" (SMS), an idea first advanced by an economist (Circacy-Wantrup, 1952), developed later by another economist (Bishop, 1978) and subsequently adopted, at least in part, by other scholars (e.g., Norton, 1992). To illustrate what is involved in this approach, suppose for simplicity that damages to some system or systems of natural resources can be entirely characterized by the size of their social cost and degree to which other substitute resources can compensate for the damage. These two dimensions of resources are depicted in Figure 1. The social costs of resource use (the vertical axis) range, at the limits, from zero to catastrophic. Substitutability (the horizontal axis in Figure 1) refers to the ability of human-made resources, including knowledge embedded in people, to substitute for natural resources, thus compensating for damages imposed on the natural system by human action.

Social costs in Figure 1 have two key characteristics: (1) They include all losses, currently and into the indefinite future, of goods and services that people in the present and all future generations do, and will, value. Unmarketed goods and services as well as those exchanged in markets are included. (2) The costs are *relative to* the benefits gained through unfettered market exploitation of resources. Thus, movement up the vertical axis in Figure 1 denotes rising social costs of resource use relative to the benefits of use received through the market.

The corners in Figure 1 depict four limiting cases of the two dimensions of natural resource and environmental management. These are (1) complete substitutability among resources and zero social costs (southeast corner); (2) complete lack of substitutability and zero social costs (southwest corner); (3) complete lack of substitutability and catastrophic costs (northwest corner); and (4) complete substitutability and catastrophic social costs (northeast corner).

Exploitation of resources that by a general consensus lie near the southwest

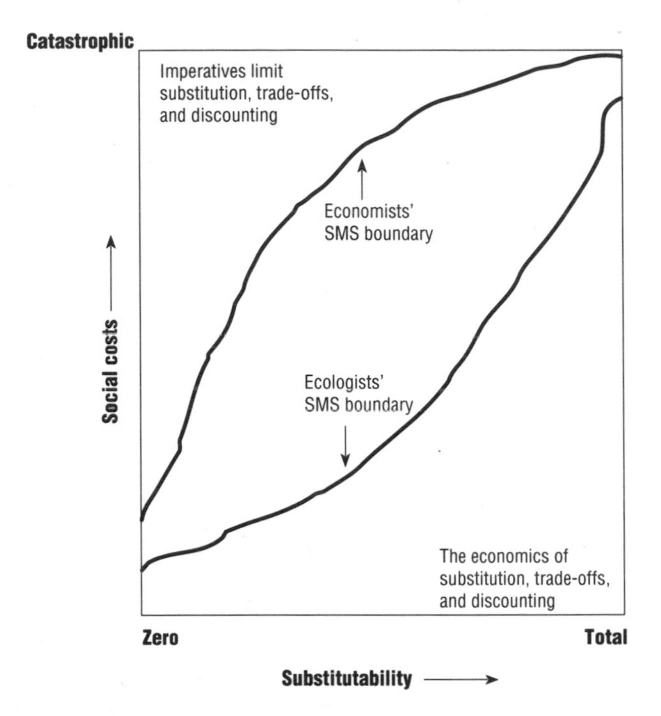

FIGURE 1 The safe minimum standard (SMS) in natural resource and environmental management. SOURCE: After Norton (1992).

or northeast corners of Figure 1 probably would excite little social concern. If the social costs of exploitation are low, then even though the resource may have few substitutes, its loss would be of little social consequence (southwest area). Exploitation of resources in the northeast area would cease well before the potentially catastrophic costs were incurred because the rise in costs, signaled either by rising prices or signs of increasing ecological stress, would induce a shift out of the resources into one or more of the readily available substitutes.

The critical area in Figure 1, the area in which potentially difficult choices have to be made about how to manage resources to assure sustainability, is an ill-defined space running from southeast to northwest. Resources which fall in the southeastern part of Figure 1 have many substitutes and low social costs of exploitation. This means that decisions about their use can be left to the play of the market where trade-offs among resources can safely occur and discounting of future costs and benefits is consistent with sustainability and intergenerational equity.

Resources for which substitutes are few and the social costs of exploitation high fall in the northwestern part of Figure 1. The closer these resources approach the northwest corner, the less management of them can be left to the play of the market or to corrective policies based on standard benefit-cost comparisons if intergenerational equity is to be respected. Under these conditions, imperatives to protect socially determined minimum stocks of the resources increasingly govern resource management, placing more severe constraints on trade-offs among resources and on discounting. In the limit, where the resource has no substitutes and additional exploitation would impose catastrophic costs, no further draw-down or degradation of the resource would be permitted.

For any particular resource or resource system the notion of the SMS comes into play when a judgment is made that the combination of limited substitution possibilities and (long-lived) social costs of exploitation is too risky to base exploitation of the resource entirely on market valuations or standard benefit-cost trade-offs. These judgments are made in the face of great uncertainty about the *physical* consequences of present and prospective exploitation (e.g., present and prospective species loss and prospects for finding substitutes for those lost) and the *social values* to be placed on the consequences. Recall that many of the resources at stake are unpriced, and the social values assigned to them must reflect current judgments about the interests of future generations.

Because of these uncertainties, judgments about when the SMS should come into play can, and do, vary widely among individuals and groups. In the end, the judgments are made through a complex social process involving the individuals and groups with contending interests in the resource or resource system. No effort is made here to describe this process, but its nature is suggested by several examples. Until recently, Americans generally were content to leave the exploitation of old-growth forests in the Pacific Northwest to the play of market forces and, for public land, the decisions of the U.S. Forest Service. But now strong efforts are made to constrain these forces to protect the spotted owl, a species seen by many to have few if any substitutes and whose extinction, and loss of associated habitat, would pose, in this perspective, high current and future social costs. In effect, a social decision has been made to shift the exploitation of the forest habitat of the spotted owl out of the southeastern part of Figure 1 into the northwestern part. Another possible example of social decisions to relocate resource management from southeast to northwest in Figure 1 is constraints put on the previously unfettered rights of farmers and land developers to destroy plant and animal habitat by draining wetlands through a mandate for "no net loss" of wetlands. In this case, however, controversy continues about what constitutes a wetland under the mandate and the extent to which different types of sites are substitutes.

The social decisions to place increasing constraints on exploitation of old-growth forests and wetlands were surrounded by enormous controversy. In terms of Figure 1, groups interested in the outcomes disagreed about the degree to which other resources could substitute for those at the center of controversy and about the social costs of unconstrained exploitation of the resources. The wavy lines in Figure 1 illustrate how these differing positions might be depicted. Because differences between economists and ecologists about these issues seem especially deep, the lines are labeled by these groups. However, this is done only to sharpen the contrasting positions. Differences about where to place particular resources or resource systems in the Figure 1 framework cut across many groups in society.

The key point made by the wavy lines is that "economists" would leave many more resources to management by the market or marketlike control policies than "ecologists" would. This is to say that for "economists" the boundary at which market (or marketlike) processes would be increasingly constrained and the SMS come increasingly into play would be farther up toward the northwest corner than the boundary that "ecologists" would draw.

One implication of the wavy lines, as drawn in Figure 1, is that "economists" and "ecologists" can agree about a class of resources that can be left to the play of the market, and about another class to which the SMS must apply. Reaching a social consensus about how to manage those two classes of resources should be relatively easy. The resources for which no management consensus exists, and which therefore pose especially difficult problems of resource and environmental policy, are those lying between the SMS boundaries of the two groups. The difference between the SMS boundaries in Figure 1 reflects different judgments by "economists" and "ecologists" about both the technical possibilities for substitution among resources and the social values that should be attached to resource losses. Most Americans probably would agree that the scenic grandeur of the Grand Canyon has few if any substitutes and that the social costs of fundamentally altering it, say by building a major dam between Lake Powell and Lake Mead, would far exceed any likely benefits. A strong consensus would place the Grand Canyon in the area northwest of the "economists" SMS boundary. However, controversy continues about the spotted owl, wetlands, climate change, and many other instances of resource exploitation.

Because the degree of substitutability among resources is in good measure a technical matter (it also has an economic component), advances in knowledge about substitution possibilities should help to narrow differences on this score. However, there are probably limits to this: adequate knowledge about long-lived, large-scale effects like climate change may not be available before the die is cast. Differences about the values to attach to resource losses are more intractable. Two people may come to agreement about the potential for other forms of social capital to compensate for extinction of the spotted owl and destruction of its habitat, while disagreeing profoundly about the social costs. (Note the vertical differences between the "economists" and "ecologists" SMS boundaries for any given degree of substitutability.) The processes by which people form the values they hold, and by which values change over time, are poorly understood.

Yet value formation does not appear to be a wholly inscrutable process. Evidence suggests that values do not emerge willy-nilly but are shaped, at least in part, by people observing the consequences of acting on them. That is, values seem to be shaped to some extent by processes of personal and social learning. This suggests that as people jointly observe the consequences of exploiting resources in certain ways, a process of social learning may, over time, narrow differences among them in the values they assign to the consequences. The decline in the number of Americans who smoke cigarettes—the increasing number who agree smoking is not a good idea—seems clearly to reflect a process of social learning of this sort.

This line of argument suggests that if we are to succeed in shaping policies to ensure sustainable management of natural and ecological resources, we badly need more information about the possibilities for substitution of human-made resources for those represented by the natural system and about the processes by which human values are shaped over time. There is a challenge here for joint work among economists and other social scientists, ecologists, and philosophers. It is a daunting challenge. Responding to it also has promise of high payoff in betterment of the human condition.

REFERENCES

Bishop, R. D. 1978. Endangered species and uncertainty: The economics of the safe minimum standard. American Journal of Agricultural Economics 60(1)(February):10–18.

Circacy-Wantrup, S. V. 1952. Resource Conservation. Berkeley, Calif.: University of California Press.

Daly, H., and J. Cobb. 1989. For the Common Good. Boston, Mass.: Beacon Press.

Ehrlich, P. R., and A. H. Ehrlich. 1990. The Population Explosion. New York: Simon and Schuster.

Norton, B. 1992. Sustainability, human welfare and ecosystem health. Environmental Values 1(2)(summer):97–111.

Simon, J. 1981. The Ultimate Resource. Princeton, N.J.: Princeton University Press.

World Commission on Environment and Development. 1987. Our Common Future. New York: Oxford University Press.

The Greening of Industrial Ecosystems. 1994.
Pp. 98–107. Washington, DC:
National Academy Press.

From Voluntary to Regulatory Pollution Prevention

FREDERICK R. ANDERSON

In environmental circles the 1990s will almost certainly bc callcd the Decade of Pollution Prevention. After 20 years of "end-of-pipe" controls on releases to the environment, attention has shifted to the elimination of potential pollution at its source. This approach, an important dimension of Design for Environment (DFE) discussed by Allenby in this volume, spans the distance from engineers attempting to redesign products and processes, to government officials seeking new incentives to encourage pollution reduction, to heads of state negotiating pollution prevention principles in international agreements based on sustainable development.

Part of the attractiveness of pollution prevention rests on the perception that it allows all interested parties to achieve their separate goals. Environmentally friendly moves by industry, on one hand, and government's willingness to employ noncoercive pollution prevention methods, on the other, have created the perception of a new cooperative approach to environmental protection. Business's optimism in embracing pollution prevention is not unexpected, particularly in light of its continuing success with energy conservation and phaseout of ozone-depleting chemicals (Anderson, 1981).

Additionally, industry benefits because pollution prevention means elimination of waste, profitable innovation, and avoidance of command-and-control regulation. By changing production processes to eliminate nonessential uses of hazardous materials, by recycling such materials, and by redesigning products to allow for reuse and for disassembly, industry can save money on production and disposal costs, waste site cleanups, and tort liability.

In addition, pollution prevention has provided industry with a new environmental consciousness. Good corporate citizenship provides valuable marketing and recruitment tools. For a more permissive philosophy to succeed, industry needs a greater measure of trust between itself and the constituencies traditionally seen in opposition—environmentalists, government, and the public. Environmentalists, regulators, and representatives of industry have already been working together on specific environmental projects such as ozone protection. Corporate leaders now seek advice on environmental issues directly from environmentalists. Representatives of environmental groups are invited to participate in industry conferences on environmental issues—a detente that would have been inconceivable a few years ago.

Government and the public benefit from business's enthusiasm for pollution prevention. Pollution prevention tends to reduce the cross-media pollution that old-style regulation sometimes produced. To the extent that industry undertakes most pollution prevention measures voluntarily, the costs of regulation can be reduced. If pollution prevention is cost-effective to industry, it means lower prices for the public as consumers.

Not surprisingly, environmentalists have also embraced the pollution prevention philosophy. To many, pollution prevention offers an opportunity to incorporate environmental values into the world economy. Greenpeace, for example, has been enthusiastically promoting its hope for a universal goal of zero discharge to the environment. The concept of sustainable development, basic to the United Nations Conference on Environment and Development, depends in part on the pollution prevention philosophy.

Pollution prevention reinforces another recent major development in environmental policy. For several years, the ranking of risks in order of priority has received major attention from the U.S. Environmental Protection Agency (EPA) leadership and the scientific community. Beginning in William Ruckelshaus's second term as EPA administrator from 1983 to 1985, the EPA leadership and EPA's influential Science Advisory Board began urging that the agency direct its energies toward the most serious environmental risks. Environmental problems are to be ranked based their relative risks and addressed in rank order by the most cost-effective means. In an interesting display of the difference between the scientific and regulatory mentalities, the EPA Science Advisory Board has urged EPA to take the initiative and address the most serious risks first, "whether or not the Agency action is required specifically by law." The board emphasized that EPA should consider pollution prevention its "preferred option" in reducing risk as "a far cheaper, more effective way to reduce environmental risk, especially over the long term" (Blomquist, 1991). Because pollution prevention strategies emphasize voluntary measures—measures generally not required under existing law—they have became the linchpin of the approach to ranking risk priorities.

Pollution prevention programs already have demonstrated significant achievements. EPA points to the enthusiastic response to its voluntary "33/50" industrial toxics project, noting that "over 400 companies have committed to an average reduction of 50 percent by 1995, for an overall reduction commitment of 335 million pounds" (EPA, 1991b). Many private entities have claimed great successes in pollution prevention. The National Electrical Manufacturers Association reports that member companies that manufacture batteries have voluntarily reduced mercury used in their production 92 percent, from 778 tons in 1984 to a projected 62 tons in 1990 (EPA, 1991a). EPA has identified a broad range of industrial success stories among individual companies with pollution prevention programs and goals. In view of such impressive statistics, it is not surprising that many are prophesying the end of the era of adversarial environmental regulation. The presumed trade-offs between economic and environmental values appear to have evaporated. Conflicts have yielded to sustainable development and other promises of benefit to both environmental and economic interests.

POLLUTION PREVENTION: A CRITICAL REAPPRAISAL

Understanding how the current enthusiasm for pollution prevention may be overstated requires drawing upon economics, law, and policy analysis of how governmental programs work in modern American society. Because of industry's enthusiasm for pollution prevention, it is appropriate to begin this critical reappraisal in the field of economics. The best economic thinking has been applied to global sustainable development, not to individual company behavior. When welfare (micro) economists begin to examine pollution prevention seriously, they will not be pleased, for two fundamental reasons.

First, pollution prevention taken to its "zero discharge" extreme contradicts the theory that pollution should be assigned a price that accurately reflects its extramarket, or "external," injury to the environment. Once pollution "externalities" are "internalized" and included in prices, a company should be permitted to pay the price and pollute. Such pollution is more economically efficient, welfare economists argue, because some industrial use of toxic substances and some discharges to the environment are valuable to society. Failure to permit these uses or discharges imposes a high economic cost and thus impedes realization of other social goals. Pollution prevention in practice tends to set as an ultimate objective zero use of, or zero discharge from, an enormous, ill-defined collection of substances and processes, without regard to economic efficiency and the high economic value of some types of pollution. Entire categories of substances and processes may be, in effect, declared "off limits" to commerce or are designated for totally enclosed systems, without regard to whether optimal discharge levels should be permitted instead, or what alternatives are therefore required.

Second, even if large numbers of substances and processes should be declared off limits to normal economic forces, advocates of pollution prevention have yet to develop an economically sound reason for industry to engage in pollution prevention over and beyond what is required by law. Why should a company voluntarily incur higher marginal production costs than a competitor incurs, merely to implement pollution prevention measures? The answers suggested by some of the arguments for pollution prevention—public goodwill, a new cooperative era, avoidance of regulation, curtailment of waste—do not seem substantial enough to carry pollution prevention for long.

The real answer appears to be profit. The initial enthusiasm for pollution prevention has developed in large part because pollution prevention is seen to be profitable to companies that invest in it. The first pollution prevention activities have in fact proven relatively inexpensive to implement, and some yield relatively high rewards—the harvest of low-hanging fruit. Industry can pick from a variety of economically attractive pollution prevention options. Yet can pollution prevention continue *indefinitely* to yield high economic returns? To answer this question, one begins in the field of economics but moves to regulatory policy analysis for a definitive answer.

Historically, the environment has supplied the least expensive factors of production—natural resources and free waste receptacles. As these resources are consumed, and the disposal of wastes increases in cost, pollution prevention becomes economically attractive in large part because of the scarcity created by regulatory measures that place some dangerous materials and disposal options artificially off limits to the economy.

As industry becomes more frugal and waste-free, opportunities for trimming waste inevitably rise in cost, and the market incentive to reduce pollution diminishes. "Technological innovation" is not a complete solution to rising costs. While current innovations in industrial production tend to be inherently less polluting, there is no technological reason why "modernization" should always produce cleaner production processes. Innovation also pollutes, although perhaps not in the same manner as an outdated technology. New technology may also produce "disbenefits" such as flimsier, less attractive, less reliable, or more dangerous products. Economists and engineers point out that optimizing for environmental protection *alone* means some loss in safety, efficiency, durability, convenience, attractiveness, or price. The proliferation of lightweight, energy-efficient vehicles has meant some loss in safety as well as comfort. Reducing CO_2 emissions by increasing the use of nuclear power raises questions relating to long-term natural resources use, safety, and waste generation.

Because pollution prevention carried to extremes flies in the face of sound economic analysis, ensuring its long-term success will be quite difficult. Replacement of polluting and dangerous substances, products, and processes will occur over the long run only if inventors and companies can rely on certain conditions:

- Substances and activities declared off limits by regulation remain so unambiguously and permanently.

• Choices about products and processes are left to market incentives, with no arbitrary regulatory specification of replacement materials and technologies.

• Regulatory decisions about which substances and activities will be "off limits" do not arbitrarily deprive the economy of useful resources (because ill-considered choices will probably be undone later, undermining earlier capital investment).

• Regulation does not take away credit for pollution reductions undertaken ahead of the regulatory schedule to give industry decision-making flexibility.

Can industry safely rely on our legislatures and regulatory agencies to abide by these ground rules? History indicates, in part, that government has powerful incentives to inject itself into almost every phase of industrial planning. Industry cannot expect legislatures and agencies to refrain from trying to manage, hasten, and order particular results. Only if industry, government, and even environmental groups work hard to counteract this tendency can pollution prevention succeed. We must all have the patience to focus on the long run.

Perhaps an example will help explain the regulatory temptation and the industry dilemma. The pollution prevention philosophy suggests that it is possible to phase out or phase down the use of the 189 toxic chemicals targeted by Section 112 of the 1990 amendments to the Clean Air Act. Indeed, one prominent EPA pollution prevention program encourages companies to attempt this task, ahead of the mandatory rollbacks that the statute will require later. Yet can industry rely on the agency, and ultimately the Congress, not to seek to control the complex process of technical innovation necessary to eliminate or drastically reduce use of these 189 economically important substances? So far regulation seems destined to thwart prevention of air toxics pollution by changing ban and phasedown objectives, mandating particular technologies, changing the rules about credit for early compliance, setting arbitrary deadlines for creative innovation, and generally seeking a counterproductive role as industrial planner, which exceeds the earlier excesses of end-of-stack regulation.

As pollution prevention costs increase, standards for evaluating performance in pollution prevention play a more important role. When large reductions in pollution are easy, everyone can afford to be lenient about how a baseline is measured or how different methods of pollution prevention are compared. As the easy reductions play out, that leniency fades. As competition heats up, the certainty, predictability, and evenhandedness of pollution reduction requirements become centrally important. Regulation will have to become far better at providing certainty, predictability, and evenhandedness. Present pollution prevention agreements negotiated company-by-company thus will become increasingly unpopular.

Faced with diminishing returns from pollution prevention, a company might consider several options. It could cease pollution prevention efforts at the level that optimizes its profits. It could seek to cooperate with its competitors, sharing technology so that everyone moves forward in unison to avoid creating competitive disadvantages within the industry. Finally, it could seek governmental implementation of uniform pollution prevention requirements, so that everyone would be subject to the same pollution prevention burdens.

If industry decides to discontinue pollution prevention, pollution prevention has ceased to be an ideal solution. Such an action would certainly fuel pressure from the public and from environmental groups to return to a primary strategy of command-and-control regulation. To these critics, industry would appear unequal to the task of self-regulation in environmental matters.

Alternatively, if companies pool their knowledge and seek to act cooperatively, they may become vulnerable to antitrust actions. So far we have heard only friendly talk about existing collaborations, but such talk may have ominous overtones when viewed from the perspective of a disgruntled competitor or the Justice Department. The EPA newsletter on pollution prevention reported efforts to develop a pollution prevention technical information exchange team among several major manufacturers. A team of representatives from these manufacturers toured one of the participating companies, facilities to observe pollution prevention initiatives. One tour member exclaimed, "It is hard to believe we are touring [this company's] facility—a major competitor!" A representative of the host company similarly remarked that "even though we compete with our products, this is an area where we must work together to ensure a safe environment." Another team member opined that "these meetings have set a precedent—never before has there been a reason so important that we could remove the barriers of competition to work toward a common goal. Good pollution prevention technology is simply too important to all of us not to share" (Li, 1991). Such sentiments possess considerable practical appeal. Perhaps pollution prevention technology should be exempted from the antitrust laws. But cooperation to eliminate competition at present is against the law.

Thus we come to the last option open to the company. It is, in fact, the last hope for the long-term future of pollution prevention. A level playing field among companies undertaking (or failing to undertake) pollution prevention appears indispensable. Government's role is to provide that level field. At the same time, government must give industry freedom to innovate, substitute products, test, market, and bring the full arsenal of technology to bear in creating a safer, more environmentally benign economy.

Such a role for government seems at present unlikely to prevail over the long run. We are too accustomed to what Professor James Krier dubbed more than 20 years ago "The Great American Regulatory Tradition." Pollution prevention begins with government encouragement of private innovation in partnership with industry to eliminate or reduce use of environmentally suspect substances and processes. But it risks metamorphosis into a new theory to justify a greater governmental role in specifying exactly what industry must do to satisfy environmental imperatives. In short, impatient regulators will not be able to keep their hands off industry's attempts to phase out or reduce use of environmentally undesirable products and production processes.

For two decades we have sought, with little success, to solve the problem of the inefficiency of uniform government-issued environmental standards. The uncritical acceptance of pollution prevention seems at present destined to guarantee that voluntary pollution prevention will be gradually supplanted by traditional command-and-control regulation that imposes on industry a tremendous variety of measures that intrude deeply into the industrial economy. Ironically, as a rationale for command-and-control regulation, pollution prevention may stifle the economy far more effectively than any social regulation of the past. In fact, as the next section shows, the process clearly has begun.

POLLUTION PREVENTION AS A NEW REGULATORY TECHNIQUE

While government's role in pollution prevention currently appears to emphasize awards, university courses, research programs, and voluntary goals and deadlines, "regulatory" pollution prevention has also clearly made an impressive debut, including substance bans, technology specifications, and incorporation of pollution prevention obligations in actions related to permits and enforcement.

While this paper focuses on federal pollution prevention, it is important to stress that the states have been at the forefront of pollution prevention activity over recent years. Some states' pollution prevention laws are simply broad policy statements in favor of pollution prevention generally. Yet other states establish specific source reduction goals and mechanisms aimed at achieving these goals. Various states require facility planning, including the submission by individual facilities of pollution prevention information, the establishment of pollution prevention plans for such facilities, submission of progress reports, and penalties or other negative incentives for failure to comply with such requirements. Some states prohibit the use of particular materials (for example, lead or other heavy metals in packaging, or CFCs in particular nonessential uses). I have dealt with these state programs in a prior paper (Moorman and Anderson, 1992).

Federal environmental laws contain a variety of potential hooks from which pollution prevention regulations and other requirements can be suspended. To enable regulation, EPA possesses a tremendous information-gathering capacity that can be used to advance pollution prevention. EPA has already indicated repeatedly its intention to make use of "traditional regulation and enforcement" in promoting pollution prevention, and has announced a search for environmental laws that it can use "to further pollution prevention within existing programs and to determine how to implement such provisions in the future" (EPA, 1991a).

The Clean Air Act offers a particularly interesting example of how the pollution prevention approach can be applied under a traditional federal environmental statute. The 1990 Clean Air Act amendments (CAA amendments) specifically incorporate pollution prevention goals (see 42 U.S.C.A. §7401(C)). The amendments provide extended compliance periods for industries that reduce their emissions of hazardous air pollutants ahead of forthcoming maximum achievable control technology (MACT) requirements. In addition, the amendments require the agency to consider pollutant source reductions, process changes, materials substitutions, cross-media impacts, and other production modifications that could reduce or eliminate emissions in setting emissions standards for hazardous air pollutants. The agency has prepared a pollution prevention guidance document for its staff working on MACT control standards for the 189 air toxics targeted by the amendments. The amendments require the phaseout of chlorofluorocarbons (CFCs) and halons and ban the use of unsafe substitutes for these chemicals. They also provide EPA with authority to regulate gasoline additives. EPA has used this authority to phase out leaded additives. EPA thus has been busily "incorporating pollution prevention into [the] regulatory process and into targeted Clean Air Act regulations" (EPA, 1991b).

Regulatory pollution prevention has not been limited to mere expansion of traditional pollution control. The EPA also intends to use the pollution prevention strategy in attacking particular pollutants in different media simultaneously. In the past, EPA has sought to control pollution within carefully defined statutory authority. Now, however, its multimedia approach means that it starts by targeting a specific pollutant then searches for various authorities with which to control that substance in all the media. For example, EPA is using a variety of statutory and regulatory tools to reduce or eliminate the source of lead (EPA, 1991a). For bans and limits based on pollution prevention, both the Federal Insecticide, Fungicide, and Rodenticide Act (FIFRA) and the Toxic Substances Control Act (TSCA) provide EPA with particularly broad authority to restrict the use of any chemical or pesticide that poses an unreasonable risk.

By combining its multimedia approach with its strategy of targeting particular industries or groups of industries for source reduction, EPA has laid the foundation for the most comprehensive (and perhaps the most chilling) regulatory application yet of the pollution prevention approach. Citing the 1990 Pollution Prevention Act, which requires the EPA to review all regulations of the agency prior and subsequent to their proposal to determine their effect on source reduction, the EPA has created its Regulatory Targeting Project covering proposed rulemaking for all media affected by 17 major industries. Under this approach future rules and permits will contain pollution prevention measures wherever possible.

EPA's approach to rulemaking for the pulp and paper industry exemplifies regulatory pollution prevention. Regulation of this industry cluster entails mandating pollution prevention through the use of stringent effluent or emissions standards that assume particular methods of source reduction. Pollution associated with this cluster will be prevented, at least in part, by end-of-pipe controls that are to be "based largely upon manufacturing processes and waste treatment systems that adopt no- or low-pollution practices" (EPA, 1991a). Rather than pollution prevention serving as an alternative or supplement to traditional regulation, it becomes the basis for regulatory standard setting.

Pollution prevention is also being used to guide the writing of permits, enforcement decrees, settlements, and money penalty reduction agreements. By incorporating pollution prevention into permitting and enforcement, EPA can achieve pollution prevention goals late in the regulatory process and, to an extent, by the back door (EPA, 1991a). EPA reports that its basic National Pollutant Discharge Elimination System (NPDES) training course now includes training on pollution prevention "to encourage and challenge permit writers to incorporate prevention into NPDES permits" (EPA, 1991b). Similarly, EPA's Office of Solid Waste is working on Resource Conservation and Recovery Act (RCRA) orientation materials "to acquaint its inspectors and permit writers with waste minimization requirements, and considering requirements for waste minimization plans as a condition for issuing RCRA treatment, storage, and disposal permits."

EPA incorporates pollution prevention requirements through its enforcement actions as well. The agency has made clear that it will include pollution prevention conditions in agency enforcement settlements. An early 1992 report detailing achievements in this area leaves no doubt about where the agency is headed (EPA, 1992). Examples of incorporating pollution prevention terms in settlement agreements during 1990 include a $42,000 reduction of a fine in exchange for an agreement to install equipment that would reduce particular wastes by 500,000 pounds per year and increase recycling of other materials by roughly 250,000 pounds per year; a reduction of a penalty by $31,000 in exchange for a binding commitment to implement a leak detection and repair program and to install in-process recycling equipment to reduce generation of 1,1,1-trichloroethane and dichloromethane; a $10,000 reduction in a fine in exchange for implementation of process changes that virtually eliminated the use of acetone in the manufacturing process in question (EPA, 1991b).

Incorporation of pollution prevention through permitting and enforcement is clearly an effective means for EPA to mandate particular pollution prevention methods or standards. Both permitting and enforcement place EPA in a position of considerable bargaining power. The application of pollution prevention strategies in such situations can be tailored to the individual circumstances of the facility or company involved. EPA might argue that permitting and enforcement allow it the flexibility to require only what is reasonable from a particular facility. Nevertheless, such tailoring is essentially at EPA's discretion and removes the game from the level playing field that is indispensable to the long-term success of the pollution prevention approach. Further, although much of the pollution prevention optimism focuses on industrywide efforts to develop goals and strategies, in permitting and enforcement individual companies or facilities may be left on their own. Although an industrywide coalition may have substantial bargaining power in discussing pollution prevention goals with EPA, an individual facility or company has considerably less stature. In sum, where broad prosecutorial and permitting discretion is guided by a sophisticated set of pollution prevention objectives, statutory authority may not be a necessary prerequisite to detailed government regulation.

CONCLUSION

Pollution prevention as regulation also appears here to stay, raising concerns about the unanticipated rapid evolution of an intrusive regulatory system. The entanglement of pollution prevention with regulation almost inevitably will limit the flexibility that industry is hoping to find in nonadversarial pollution prevention. Thus industry, government, and environmental groups must critically examine the inner workings of pollution prevention and be vigilant to deter it from becoming a rationale for national industrial policy and a more intrusive form of regulation.

REFERENCES

Anderson, Frederick R. 1981. A Connecticut Yankee in King Energy's Court. ABA Journal 67(June):722.

Blomquist, Robert F. 1991. The EPA Science Advisory Board's report on "Reducing Risk": Some overarching observations regarding the public interest. Environmental Law 22:149.

Environmental Protection Agency. 1991a. Pollution Prevention 1991: Progress on Reducing Industrial Pollutants. Washington, D.C.: U.S. Environmental Protection Agency, Office of Pollution Prevention.

Environmental Protection Agency. 1991b. U.S. EPA Progress in Meeting Congressional Mandates of Pollution Prevention. Washington, D.C.: Environmental Protection Agency, Office of Pollution Prevention.

Environmental Protection Agency. 1992. Pollution Prevention Through Compliance and Enforcement. January. Washington, D.C.: Environmental Protection Agency, Office of Pesticides and Toxic Substances.

Li, Phil. 1991. Aerospace industry establishes prevention exchange team. Pollution Prevention News (December):3. Washington, D.C.: Environmental Protection Agency, Office of Pollution Prevention.

Moorman, James W., and Frederick R. Anderson. 1992. The regulatory future of pollution prevention. In Proceedings of the 2nd Annual Environmental Technology Exposition and Conference, Washington, D.C., 7–9 April 1992.

The Greening of Industrial Ecosystems. 1994.
Pp. 108–122. Washington, DC:
National Academy Press.

International Environmental Law and Industrial Ecology

ROBERT F. HOUSMAN

If sustainable development is ever to be achieved, the course of industrial development must be charted to serve not only concerns for economic competitiveness, but also concerns for an ecologically sustainable future. If industrial ecology is to provide a common ground between the industrial and environmental agendas, it is important to consider the impact of international environmental law on industrial ecosystems and what steps might be taken to effect needed systemwide changes.

Briefly summarized, industrial ecology represents a systemwide approach to analyzing industrial processes. With this approach it is possible to evaluate how environmental concerns and costs may be integrated into industrial and economic decision making, and to maximize the beneficial use of resources while minimizing disruptions to the industrial ecosystem.

In some cases, the integration of environmental factors into business decisions has resulted in a direct cost-savings to the manufacturer, creating an economic incentive sufficient to encourage the adoption of these principles. However, in the majority of cases, such efforts will not be cost-effective either in the short term, because of start-up costs, or in the long term, because market failures may prevent true environmental costs from being included in the cost-benefit analysis. In this context, law can play a critical role, since it embodies a host of noneconomic social value judgments (e.g., saving a species is good). Laws, therefore, can be crafted to correct market failures and provide incentives for undertaking activities that are cost-effective only over the long term.

At the national level, particularly in the United States and other industrialized countries, domestic environmental laws are slowly shifting to providing incentives to protect the environment. For example, the United States in 1990 adopted the Pollution Prevention Act (42 U.S.C. §13101 (a)(3)), which explicitly recognizes that

> There are significant opportunities for industry to reduce or prevent pollution at the source through cost-effective changes in production, operation, and raw materials use. Such changes offer industry substantial savings in reduced raw material, pollution control, and liability costs as well as help protect the environment and reduce risks to worker health and safety.

The act also recognizes that current environmental laws often impede cleaner production and consumption practices. To remedy this, the act declares a national policy to eliminate these disincentives and, to the greatest extent feasible, provide incentives for the wider acceptance of pollution prevention techniques for environmental protection. Similarly, the new German auto initiative sets standards for reuse and recycling of used automobiles. Domestic laws such as these seek to internalize environmental costs in decision making and to encourage consideration of the environmental impacts of resource use.

INDUSTRIAL ECOLOGY AND INTERNATIONAL ENVIRONMENTAL LAW

The vast majority of international environmental law focuses on pollution abatement, remediation, and compensation for the harms stemming from environmentally hazardous activities. In essence, the international system of environmental law has adopted the traditional command-and-control approach to environmental regulation entrenched in the domestic environmental laws of most nations. Because the international legal system traditionally lags behind developments in the domestic laws of the respective nations, the movement toward preventing environmental harm remains on the distant horizon of international law. Of course, the relative rate at which pollution prevention and cleaner production concepts are incorporated into international law will largely depend on both the rate at which these concepts receive wider incorporation into the domestic laws of nation-states and the wealth of exogenous factors that shape international affairs and law generally. This lack of attention to elements of improving the environmental characteristics of industrial ecosystems—encouraging pollution prevention, reuse of resources, and the systems aspects of cleaner production—in current international environmental law is troubling in at least two respects.

First, with the increasing globalization of economic development at the regional and multilateral level—with the increasing emphasis on free trade and free trade agreements—there is increasing demand for international environmental agreements. As the field of international environmental law continues to grow, it will play a major role in determining how environmental protection priorities are developed. It is therefore important to ensure that developments in international environmental law shift to providing incentives for systemwide reduction of pollution and efficient use of resources. Absent such measures, there is the increased risk that these newly emerging laws may hinder the more efficient use and reuse of resources at the global level.

Second, as the globalization of economic development continues to expand, international environmental laws not only must avoid disincentives to improving the environmental characteristics of entire industrial ecosystems, but must provide an impetus for the adoption of these systems-based approaches. The external factors that hinder the broader adoption of pollution prevention, frugal resource use, and a systems approach to environmental issues at the national level are even more numerous and insidious at the international level. Therefore, law must serve a larger role in encouraging those environmentally preferable practices.

MOVING BEYOND "COMMAND-AND-CONTROL" IN INTERNATIONAL ENVIRONMENTAL LAW

International environmental law, like many national domestic environmental laws, is directed primarily to controlling pollution at the end of processes. A more comprehensive approach is needed to encourage systemwide changes in complex production and consumption practices. A wide range of regulatory approaches and devices could be incorporated to encourage systems approaches to addressing environmental concerns in international environmental law. A number of these approaches are discussed below; however, the list laid out here is in no way exhaustive. Moreover, the following approaches are not necessarily mutually exclusive, and the best regulatory approach may consist of a combination of these, as well as other potential approaches.

Preference for Market Structure and Market-Based Approaches

There are four general approaches to environmental law and regulation: (i) liability schemes; (ii) command-and-control; (iii) market structure strategies (strategies for developing or defining a market, such as take-back legislation); and (iv) market-based strategies. At the international level, states are liable for harm from transboundary pollution under the doctrine of state responsibility. However, the system is cumbersome and little used. Moreover, it does not appear to be the leading contender for further international development to address global environmental issues. The second, and most widely used, approach to environmental law and regulation at both the national and the international level is the command-and-control approach. The principal benefit of such strategies is apparent certainty (although they often bring with them many unintended and uncertain consequences). Because of the potential certainty inherent in command-and-control strategies, they remain critical for addressing environmental problems with irreversibilities, such as acute toxic pollution and loss of species. This certainty has made this approach to environmental regulation the strategy of choice for many, if not most, nongovernmental environmental organizations, who remain suspicious of industry's commitment to environmental protection. Increasingly, however, it is recognized that we have reached the limits of command-and-control regulations and must move toward more market-oriented strategies (i.e., market structure and market-based strategies) for efficiently capturing the next needed increment of environmental protection.

The economic argument for market-oriented strategies, such as taxes and tradable emission permits, is that they can achieve the same level of environmental protection provided by command-and-control regulatory approaches but at less expense. This is because market-based strategies in particular provide the flexibility for firms to select the lowest-cost method of achieving the selected level of environmental protection. Moreover, the flexibility provided by such strategies also encourages technological innovation.

There is, however, a view to the contrary among some environmentalists, namely, that market-oriented approaches to environmental protection are a furtive way to roll back environmental standards. Attacking command-and-control strategies, or liability schemes (as tempting as that surely is for many), therefore, may not be the best approach to shifting the regulatory focus to market-oriented approaches. Rather, such a regulatory shift may be most easily achieved through cooperation between regulators and environmentalists to demonstrate the benefits of market-oriented strategies and to develop appropriate confidence-building mechanisms as a prelude to a shift from command-and-control strategies.

Negotiation and Other Confidence-Building Efforts

Negotiation and other confidence-building mechanisms can take place either within the context of the regulatory process among government, industry, and environmentalists (such as the Environmental Protection Agency's [EPA's] "reg-neg," or negotiated regulation, approach to the regulations under the 1990 Clean Air Act Amendments), or outside the regulatory process (such as the Environmental Defense Fund's efforts to assist the McDonalds fast-food chain with its waste reduction program).[1]

Because of the relative lack of international environmental regulations, negotiation and confidence-building efforts are particularly important at the international level to fill the void left by law. Moreover, because such efforts provide for a degree of innovation, flexibility, and less politicized decision making not usually found in the context of international agreements or national laws, they can be an important impetus for furthering environmental protection efforts.

Regardless of one's opinion of their results, past efforts at confidence building provide some guidance about what elements are necessary for such efforts to be successful. The process must be open to all parties that make the effort to participate. The process must also be transparent—that is, all the materials used in, and stemming from, the process must be available publicly. To avoid even the appearance of collusion, all parties must be able to participate in such exercises on an equal footing with industry and government. Thus, at times financial and technical support will have to be provided to the environmental organizations. Further, to develop long-term credibility, monitoring mechanisms must be built into negotiated agreements.

Credibility is perhaps the most important ingredient in the confidence-building process. If governments or industries are perceived as being disingenuous, confidence-building mechanisms will prove futile. Similarly, if environmentalists are perceived as being incapable of dealing in good faith, industry will refuse to work with them. Credibility involves many elements, on all sides of the negotiating table, with regard to the presentation of commitments, positions, rhetoric, and "facts." To a great extent, the credibility of the parties will be determined by the results of their joint efforts. For example, if a market-based strategy is used in an inappropriate situation and environmental harm results, credibility will be lost even if the strategy was a product of negotiated decision making. If, however, a negotiated approach results in both environmental and economic gain—a "win-win" situation—then credibility levels will increase accordingly. Thus, if industry wants to develop trust among environmentalists, it must ensure that joint efforts result in real environmental benefits.

Finally, industry and environmentalists must be pragmatic about what they think they can achieve through negotiation. Industry must realize that the international environmental community is not of one mind—there will always be a watchdog watching the watchdog—and dissension to almost any settlement is likely from some segment of the environmental community. Industry must also realize that the confidence it seeks from such exercises does not derive from the unanimous adoption of the result but from the insulation provided by the open process itself. Similarly, environmentalists must recognize that negotiation is a process of compromise, which requires them to develop their goals and set priorities among them. Moreover, both must recognize that, even if a compromise does not result from a particular negotiation effort, if all parties have participated openly and honestly, confidence levels will increase and assist in future endeavors. Finally, industry and environmentalists must realize that confidence-building negotiation exercises are not a replacement for participatory governmental and intergovernmental processes. They are, however, a vital component of forward-thinking environmental protection strategies.

Internalizing Environmental Costs

Ideally it would be desirable to have all external environmental costs be fully internalized into markets through proper pricing of environmental values, which are now largely free or otherwise underpriced. This ideal cannot, of course, be achieved in practice, but the need for cost internalization is generally recognized. In the broader context of sustainable development, the need for cost internalization is recognized and encouraged through the "polluter pays" principle. The United States and other countries in the Organization for Economic Cooperation and Development (OECD) already accept this principle, in theory at least, although implementation through law and regulation at national and international levels is sporadic, weak, and generally inefficient.[2] Thus, environmental improvements in industrial ecosystems could be facilitated globally through wider acceptance, and improved implementation, of the polluter-pays principle at the international level.

One method for internalizing environmental costs that is now being considered in the context of efforts to halt global warming is to sell limited numbers of marketable pollution rights, or permits, on open markets. This method is particularly appealing internationally with regard to resources generally thought of as part of the global commons (e.g., the oceans and the atmosphere). While pollution rights raise interesting questions of environmental ethics and equity, such rights do force companies to include at least some artificial measure of their environmental costs as real costs, and thus provide an incentive for companies to reduce them.

An alternative method of encouraging international internalization of environmental costs is to allow national governments to place countervailing duties on imported products equal to the production subsidy the goods receive in the country of origin as a result of less stringent environmental regulations.[3] There are, however, a number of practical difficulties with implementing such a system.

Precautionary Principle

The precautionary principle requires that if a particular action can be shown to pose a threshold risk of harm, then the proponents of the action must prevent or terminate that activity unless they can prove by a preponderance of the evidence that the activity will not degrade the environment.[4] The precautionary principle is particularly effective at guiding decision making where little or no evidence exists as to the potential environmental risks of an action.

The precautionary principle is closely related to the principle of pollution prevention; each seeks to avoid environmental harms before they occur. The precautionary principle is also closely related to the principle of sustainable development; each maintains that humankind will confine its actions to those activities that do not cause irreparable damage to the environment. Thus, the precautionary principle provides important common ground between international environmental law and efforts to prevent environmental damage and is a natural starting point for the wider incorporation of such preventive measures into international environmental law.

It is therefore encouraging that the precautionary principle is increasingly being developed as a central tenet of international environmental law, as in Principle 15 of the Rio Declaration, recently adopted at the United Nations Conference on Environment and Development.[5] Similarly, the precautionary principle has been included in a wide number of international environmental declarations and treaties, including the 1982 World Charter for Nature,[6] the Convention for the Prevention of Marine Pollution from Land-Based Sources,[7] the Barcelona Convention,[8] the London Declaration of the Second North Sea Conference,[9] and the Draft Ministerial Declaration for the Second World Climate Conference.[10] Moreover, the precautionary principle has also been adopted in the context of a number of international economic treaties and declarations, including the negotiating text of the Treaty on European Union (Maastricht Treaty),[11] the 1991 OECD Ministers of the Environment Declaration,[12] the 1990 Economic Summit of Industrialized Nations,[13] and the 1989 Summit of the Arch.[14]

Pollution Prevention

Although international environmental law remains largely media-specific, it has shown a proclivity for pollution prevention as an environmental protection strategy. For example, the Montreal Protocol[15] ultimately contemplates a ban on chlorofluorocarbons, requiring companies to redesign their production systems and products to eliminate ozone-depleting gases. A similar approach is likely to be followed with respect to reducing global warming emissions in protocols negotiated to control the emissions. These and other efforts at the international level provide substantial support for pollution prevention or designing pollutants, to the greatest extent possible, out of the production process. Future international agreements should recognize the success of these pollution prevention schemes and should adopt, where appropriate, similar formats. Perhaps, however, these future efforts should focus greater attention on providing incentives for pollution prevention than on the command-and-control efforts discussed above.

Recycling and Reuse Requirements

While pollution prevention can assist companies to minimize their wastes, it does not address the ultimate disposition of any unavoidable wastes from the product cycle. International environmental law can play a vital role in encouraging reuse and recycling by helping to create a market for "waste." International environmental law can encourage companies to design for the environment by placing reuse and recycling requirements on products sold domestically and traded internationally. Such recycling and reuse requirements can be brought about both through international laws (e.g., agreements), and through domestic laws that apply to the product as it enters a given country's market. Reuse and recycling requirements can be achieved through a variety of legal mechanisms. For instance, law could be used to create incentives, such as lower tariff schedules for products with higher percentages of recycled content. If incentives prove unfeasible, law could place regulatory burdens, such as embargoes or higher tariffs, on products that fail to meet a minimum requirement for reused and recycled content. Law can also assist in ensuring that a sufficient supply of reusable and recyclable wastes is available to support a market for waste through mechanisms such as the deposit system in Sweden for aluminum cans.

Recycling and reuse requirements are now attracting increased international attention since the enactment of a Danish law imposing glass reuse requirements on the foreign and domestic companies selling certain products in bottles in Denmark. Unlike deposit schemes familiar in the United States as a way to encourage recycling, the Danish system is a reuse (i.e., collect, clean, refill) requirement. In the aftermath of the "Danish Bottles Bill," Germany adopted laws requiring the recycling and reuse of product packaging materials and automobile components. In response to the German legislation, the European Community is now considering similar laws requiring reuse and recycling in packaging. Similarly Ontario's environmental levy on nonreuseable bottles also provides a market incentive for reuse.

Incentives

Although much of environmental law at both the national and the international level focuses on imposing penalties against rogue actors, the traditional "penalty-oriented" approach in international environmental law has to be augmented with environmental laws that provide incentives for cleaner production and consumption practices. For example, while much of the Montreal Protocol focuses on penalties to discourage free riders and to encourage nations to join the Protocol, many of the Protocol's most powerful provisions provide incentives, including financing for the transfer of environmentally sound technologies, to encourage states both to join the Protocol and to abide by its goals. A similar approach, whereby the international community provides incentives to countries, and companies, that prevent pollution, use resources efficiently, and recycle or reuse their products might accelerate the adoption of such practices faster than penalty-oriented international legal frameworks.

Technology Transfer

The requirement to provide technology transfer to developing countries, in the context of international environmental agreements, is emerging as a new norm of international environmental law. The Montreal Protocol, which provides mechanisms for increasing developing countries' access to technologies that reduce emissions of ozone-depleting chemicals, is a primary example of this trend in international environmental law. Similarly, all agreements coming out of the 1992 United Nations Conference on the Environment and Development contain some provision for increased access to environmentally sound technologies.[16]

Technology transfer has the potential to become one of the most important incentives for further improving the environmental performance of industrial systems. Opportunities to transfer technology can provide incentives for industries to look at their production processes at the design stage, thereby encouraging pollution prevention. In turn, markets for environmentally sound technologies can encourage innovation of new technologies as well. As industrial ecology provides insights into technology innovation, technology transfer can play a vital role in broadening the application of these emerging technology innovations.

Ecosystem Approach

Another method for encouraging the wider use of practices that address systems aspects of environmental problems is for environmental treaties to adopt an ecosystem approach to environmental management. While there has been little international regulation of the manufacturing and disposal practices of market actors, where such international regulation exists it generally does not look at the environmental effects of the manufacturing cycle as it relates to the environment as a whole. Thus, for example, the MARPOL Convention,[17] which requires, among other things, that wastes generated on ships while at sea must be disposed of at port, does not control the ultimate disposal of these same wastes; once offloaded, such wastes can be dumped by harbor barges into territorial waters without regulation (unless the nation is a signatory to the London Dumping Convention,[18] a separate international agreement). The failure of environmental agreements to regulate on an ecosystem basis allows environmental harm simply to be shifted rather than addressed comprehensively.

A comprehensive ecosystem approach to international environmental law would encourage producers, intermediate users, and final users and disposers to eliminate environmental costs at the design and consumption stages. Although an ecosystem approach has been slow to develop, international environmental law is, after all, borne out of an understanding that environmental threats do not respect national boundaries. It is, thus, sympathetic to the protection of ecosystems as units, even if it continues to do so in a media-specific manner. Moreover, recent developments bode well for the broader adoption of an ecosystem approach to international environmental law. Through the use of the Global Environmental Monitoring System, and the increased use of data provided by satellite, nations are increasingly looking at cooperative efforts to address threats to ecosystems at the regional level.

Eco-Labeling

Closely related to the principle of environmental cost internalization is the concept of eco-labeling. Eco-labeling requires manufacturers to disclose the relative environmental costs attendant to the manufacturing, use, and disposal of products, whether those costs are internalized or externalized. Eco-labeling thus allows the consumer to take these environmental costs into account when purchasing a product and also encourages producers to minimize the environmental costs of their products. Eco-labeling requirements that focus on the overall disclosure of environmental costs, as opposed to minimum standards to carry a seal of approval, provide an incentive for manufacturers, and users, to reduce the environmental effects of their activities to the greatest extent possible. In fact, merely by requiring, as a preliminary step, a manufacturer to inventory the environmental costs of a product, eco-labeling makes manufacturers more conscious of the environmental waste and costs associated with their operations. This increased consciousness on the part of manufacturers can, in and of itself, encourage manufacturers to change their patterns of behavior. For example, in response to a similar inventory of toxic emissions, required by the U.S. Emergency Planning and Community Right to Know Act,[19] the Monsanto Company voluntarily committed itself to reducing its air toxic emissions by 90 percent over a four-year period.

A great deal of international attention is now focused on the development of international eco-labeling standards. For example, at the regional level, the European Community has already adopted an eco-labeling system. International eco-labeling can also be effected through an importing nation's eco-labeling requirements that apply to both domestic products and imported products. Existing unilateral labeling requirements on internationally traded products already apply to dolphin safety and use of tropical hardwoods. An across-the-board international labeling regime would, however, be cumbersome and problematic from the standpoint of industry and the regulating and oversight body. Despite such difficulties, the International Standards Organization is hard at work in developing a voluntary program for international eco-labeling.

Environmental Assessments

Environmental assessments are another legal tool to encourage the proponent of an action to take the potential environmental costs of that action into account before undertaking the action, thereby encouraging the proponent to design the action to eliminate, or mitigate, these costs. By requiring that a project's planning process take into account the potential environmental effects of the project's development, environmental assessments facilitate consideration of environmental factors during design, rather than after the fact.

While it is unclear whether international environmental law currently recognizes a duty to assess the environmental impact of one's actions, environmental assessment is rapidly emerging as a central principle of international environmental law, having been recognized by the vast majority of countries in their domestic laws, internationally in a number of different contexts,[20] and by most international organizations.[21] Additionally, under contemporary international law, and U.S. domestic law, there are few requirements that proponents of private projects prepare an environmental assessment for an action in the absence of a connection between the action and a governmental or intergovernmental entity (usually in the form of governmental or intergovernmental funding or permitting). This may, however, be changing. Mexico, for instance, has adopted an environmental assessment law that applies to both private and public actors.

Increasing the scope of international and domestic requirements for environmental assessment to apply to a greater range of private actions would make private actors more aware of the environmental costs of proposed actions. Moreover, extending the scope of international environmental assessment requirements merely recognizes that as private development initiatives continue to play a larger role in global development, unless properly designed and implemented, these initiatives can entail environmental impacts similar to those of governmental and intergovernmental development efforts. Such an extension of existing law has already been acknowledged internationally by the OECD.[22]

IMPEDIMENTS TO SYSTEMS APPROACHES IN INTERNATIONAL ENVIRONMENTAL LAW

In contrast to the relatively nascent body of international environmental law, there exists a much older and more well defined body of international trade law that can come into conflict with efforts to take a systems approach in international environmental law. The General Agreement on Tariffs and Trade[23] (GATT) is the principal international instrument whose rules govern the vast majority of international trade. A number of GATT's rules may thwart efforts to incorporate incentive-based approaches in international environmental law (including both the extraterritorial application of domestic laws and pure international law) set out above.[24]

For example, GATT generally prohibits trade restrictions based on the production process of a product. Unless a restriction applies directly to the product, in some way altering the product's characteristics, the restriction is likely to conflict with GATT. It is unclear under GATT's current rules what type of control (i.e., governing the physical or chemical composition of a product, as opposed to its ability to perform a given task) is needed for a regulation to pass muster under GATT. Thus, a requirement governing the recycled or reused content of imported or exported products might violate GATT. Moreover, requirements that labeling disclose production process methods and their relative environmental costs may violate GATT, as well. Similarly, GATT currently prohibits a country from imposing countervailing duties on imported products that are manufactured under diminished environmental protection standards.

While GATT does not prevent the international community or a nation from adopting international environmental law provisions that conflict with GATT's rules, GATT can be used by other parties to penalize the enacting nation or nations for their trade law transgressions. Additionally, since most countries currently are seeking increased free trade, international environmental laws that restrict trade in violation of GATT place their enacting nations in a most awkward position internationally.

International Inertia

The inability of the international system of law to evolve over time has prompted one scholar to liken international law to "post-feudal society cast in amber."[25] To the extent that industrial ecology implies that there is a need to rethink how international law has traditionally viewed environmental issues, it is important to realize that these changes may be difficult to realize. Thus, forward-looking ideas, particularly those affecting, directly or indirectly, what nations perceive as their sovereign right to use "their" resources (e.g., pollution rights), may prove difficult to advance as international legal frameworks. Even where change may be possible, international consensus is not easily built. Therefore, international law may be an inappropriate mechanism for advancing systems approaches to improving the environmental performance of industrial ecosystems. This produces a terrible catch-22. Markets are increasingly global, and national governments are turning over greater authority to regulate these markets to an international community that lacks the necessary legal structures and mechanisms to do so.

The Lack of an International Enforcement Mechanism

At the national level, law can be looked to as a means of encouraging, through the threat of sanctions, market actors to respond to environmental imperatives. At the international level, there is little in the way of effective enforcement of international law. Assuredly, nations can use trade sanctions to encourage foreign producers, and in turn their national governments, to adhere to environmental standards; however, these sanctions suffer from the same trade law impediments and inefficiencies discussed above. Without an effective international enforcement mechanism, there is little guarantee that foreign governments will compel their industries to, for example, internalize their full environmental costs into the costs of their products.

Finite Economic Resources

To the extent that changes in international environmental law to encourage systemwide alterations will require nations to divert already strained resources from existing priorities, such proposals may not readily come to fruition. Much of the funding necessary to provide the public financing to adopt incentive-based approaches can be obtained through tax schemes that internalize environmental costs; however, it must be remembered that tax schemes are, themselves, a form of wealth redistribution. Thus, while incentives to encourage nations and their industries to shift to cleaner production and consumption practices may be preferable over penalties for the failure to comply with these practices, if these incentives require nations to make additional fiscal resources available at the international level, such incentive-based approaches may be more problematic than the less effective penalty-based approaches. A quick look at the mired environmental funding discussion taking place in the context of the North American Free Trade Agreement illustrates this problem well.

CONCLUSION

For sustainable development to move from rhetoric to reality, industrial activity must move beyond its narrow short-term economic agenda. The energy, innovation, drive, and creativity that characterize the private industrial sector must also serve an environmental agenda. Industrial ecology is critical to harmonizing the economic and environmental agendas.

Society is now looking to engineers to help meet one of the greatest engineering and social challenges the world has ever faced—achieving sustainable development. As the role of engineers evolves to meet this challenge, so too must the role of law evolve in support of these efforts.

International environmental law can serve as an important impetus for improving the environmental performance of industrial ecosystems. To date, however, international environmental law has paid little attention to this effort. Thus, if international environmental law is to play a role in developing cleaner industrial ecosystems, then the existing tenets of international environmental law must first be modified to include, whenever appropriate, a preference for market-based, ecosystem approaches that encourage pollution prevention schemes over command-and-control and liability schemes. Fortunately, international environmental law appears to be developing, albeit slowly, in a direction that is consistent with the approach to environmental protection advocated by industrial ecologists.

ACKNOWLEDGMENT

The author wishes to thank Brad Allenby, Durwood Zaelke, Barry Pershkow, Nancy Benton, and David Hunter for their assistance. Any remaining errors or omissions are the sole fault of the author.

NOTES

1. It should be noted that EDF's efforts working with McDonalds were made possible by a more adversarial effort conducted by Lois Gibbs's group, the Citizens Clearinghouse for Hazardous Wastes.
2. OECD, The Polluter Pays Principle: Definition, Analysis and Implementation, May 26, 1972, C(72)128.

3. Robert Housman and Durwood Zaelke, *Trade, Environment and Sustainable Development: A Primer*, 15 Hastings Int'l & Comp. L. Rev. 535, 606 (1992).
4. Margaret Spring, *The Precautionary Principle: An Update*, June 18, 1992 (Center for International Environmental Law Working Paper).
5. Rio Declaration, Principle 15, *reprinted in* Reuters, June 14, 1992 (final text).
6. United Nations General Assembly, World Charter for Nature, A/37/L.4 and Add.1 at para. 11 (Oct. 28, 1992).
7. The Convention for the Prevention of Marine Pollution from Land-Based Sources, Feb. 21, 1974, 13 I.L.M. 352 (1974).
8. Sixth Meeting of the Contracting Parties to the Convention for the Protection of the Mediterranean Sea Against Pollution, Recommendations Approved by the Contracting Parties (Oct. 1989), *reprinted in* Greenpeace Paper 28 (1990).
9. Second International Conference on the Protection of the North Sea, Ministerial Declaration, at art. XVI(1) (London, England, Nov. 1987).
10. Second World Climate Conference, Geneva, Switzerland, Oct. 29-Nov. 7, 1990, Draft Ministerial Declaration, at preamble, art. 7.
11. European Union Treaty, Final Act of the Conference of Member State Representatives, signed in Maastricht, Feb. 2, 1992, art. 130R(2), *reprinted in* European Reprint, No. 1746, Feb. 22, 1992.
12. OECD Ministers of the Environment Declaration, OECD Environment Committee at Ministerial Level Communique para. 38 (Jan. 31, 1991).
13. 1990 Economic Summit of Industrialized Nations, Economic Declaration para. 62 (July 11, 1990).
14. Summit of the Arch Economic Declaration art. 34 (Paris, July 16, 1989).
15. The Montreal Protocol on Substances That Deplete the Ozone Layer, adopted and opened for signature, Sept. 16, 1987, *entered into force* Jan. 1, 1989, 26 I.L.M. 1541 (1987) *as modified by* The Montreal Protocol on Substances That Deplete the Ozone Layer, London 1990, Annexes A,B, UNEP/OzL.Pro.2/3 (1990).
16. *See* Rio Declaration, Principle 9 (reprinted in Reuters, June 14, 1992 (final text)); UNCED, Convention on Biological Diversity, June 5, 1992, UNEP/Na.92-7807, art. 16; UNEP, Non-Binding Authoritative Statement of Principles for a Global Consensus on the Management, Conservation and Sustainable Development of All Types of Forests, June 13, 1992, A/CONF.151/6/ Rev. 1, art. 11; UNCED, United Nations Framework Convention on Climate Change, A/AC.237/ 18(Part II)/Add.1 (Advance Copy-Final), May 15, 1992, art. 4(c); UNCED, Agenda 21, Chapter 33 (unchanged negotiating text reprinted by Econet).
17. International Convention for the Prevention of Pollution from Ships, 1973, *reprinted in* 12 I.L. M. 1319 (1973) *as modified by* Protocol of 1978 Relating to the International Convention for the Prevention of Pollution from Ships, 1973, *opened for signature* June 1, 1978, I.M.C.O. Doc. TSP/CONF/11 (1978), *reprinted in* 17 I.L.M. 546 (1978).
18. Convention on the Prevention of Marine Pollution by Dumping of Wastes and Other Matter, *done at* London, Dec. 29, 1972, 26 U.S.T. 2403, T.I.A.S. No. 8165, 1046 U.N.T.S. 120 (entered into force Aug. 30 1975).
19. 42 U.S.C. §§ 11001-11050.
20. *See e.g.* Lac Lanoux (Spain v. Fr.) 12 R. Int'l Arb. Awards 281, 315 (1956) (international arbitration decision including in dicta acknowledgment of duty to assess); Espoo (Finland) Convention on Environmental Impact Assessment in a Transboundary Context, ECE, June 21, 1991, 30 I.L.M. 802; Wellington Convention on the Regulation of Antarctic Mineral Resources Activities, June 2, 1988, AMR/SCM/88/78, 27 ILM 868, arts. 4, 37(7)(d)-(e), 39(2)(c), 54(3)(b) (duty to assess in a global multilateral treaty); Council Directive 85/337/E.E.C. of June 27, 1985, O.J.E.C. no. L 175/40 of July 7, 1985 (duty to assess in a regional multilateral context).
21. *See e.g.* OECD, OECD Council Recommendation: Assessment of Projects with Significant Impact on the Environment, May 8, 1979, OECD C(79)116; World Bank, World Bank Opera-

tional Manual, O.D. 4.00, Annexes A, A1,A2, Nov. 1989; UNEP, Governing Council Decision: Goals and Principles of Environmental Impact Assessment, UNEP/GC.14/17 Annex III, UNEP/GC/DEC/14/25, June 17, 1987.

22. OECD, OECD Council Recommendation: Analysis of the Environmental Consequences of Significant Public and Private Projects, Nov. 14, 1974, OECD C(74)216.
23. General Agreement on Tariffs and Trade, *opened for signature* Oct. 30, 1947, 61 Stat. A3, 55 U.N.T.S. 187.
24. *See* Robert F. Housman and Durwood J. Zaelke, *Trade, Environment, and Sustainable Development: A Primer*, 15 Hastings Int'l & Comp. L. Rev. 535, 539-41 (1992).
25. Philip Allot, International Law and International Revolution: Reconceiving the World 10 (1989).

The Greening of Industrial Ecosystems. 1994.
Pp. 123–133. Washington, DC:
National Academy Press.

Industrial Ecology: The Role of Government

MATTHEW WEINBERG, GREGORY EYRING,
JOE RAGUSO, and DAVID JENSEN

There is nothing more difficult . . . than to take the lead in introducing a new order of things.

Niccolo Machiavelli, *The Prince*

The relentless pace of technical and industrial advancement over the last century has fundamentally transformed the relationship between human society and the natural world. As the scope and range of human activities have expanded exponentially, profound and possibly irreversible environmental changes have been set in motion.[1] For the first time in history, humankind can potentially alter the basic biophysical cycles of the earth.

Modern social systems have clearly broken away from the patterns of global ecological stability that existed during the 2 million years when humans lived in small nomadic bands (Ponting, 1990). But realistically there can be no turning back. While some believe that humanity's capacity for technical and economic progress is virtually boundless, the fact that human activities are now resulting in materials flows commensurate with those of nature should give one pause. Human activities are estimated to release several times as much mercury, nickel, arsenic, and vanadium to the environment as do natural processes, and more than 300 times as much lead (Galloway, 1982; see also Ayres, 1992). Concentrations of carbon dioxide in the atmosphere are increasing at a rate 30 to 100 times faster than observed in the climatic record; methane concentrations are increasing 400 times faster than historically (U.S. Congress, Office of Technology Assessment, 1991a).

The challenges for our institutions of governance in addressing emerging en-

NOTE: The views expressed here are those of the authors alone and do not necessarily reflect those of the Office of Technology Assessment.

vironmental threats are indeed formidable. Traditional formulas of environmental management will no longer suffice. The reactive strategies of remediation and "end-of-pipe" modification will have to be supplanted by new strategies and approaches. Systematic changes are needed in materials use, production processes, product formulation, product use, and disposal practices.

The evolving concept of industrial ecology could serve as a significant catalyst in transforming societal patterns of production and consumption.[2] Although at first glance the term *industrial ecology* appears to be an oxymoron, the notion of connectivity and interdependence that it embodies is extremely important. Interdependence is the dominant phenomenon of our age. With elaborate webs of production now stretching across the globe, the economic destinies of nations are closely intertwined.[3] Since these production networks draw on the natural resource endowments of countries around the world, our economic activities, regardless of how localized they appear, are becoming more closely tied to global ecological disruption[4] (Wyckoff and Roop, 1992). Tighter economic linkages among nations are accentuating the world's environmental interdependence.

Societies can no longer separate economic imperatives from ecological imperatives. Economic productivity cannot be improved if the natural ecosystems on which the economy depends are undermined. Industrial enterprises need to create the same strong linkages at the postconsumer end of our economies that exist at the front end. Today's highly efficient one-way systems of providing goods and services must give way to circular systems of production. Use of both products and product waste streams needs to be optimized (Frosch, 1992). At all stages of production and consumption, actors must shape activities to minimize both resource use and waste generation. A true systems view of our industrial activities, and the impact of those activities on the environment, is required. This implies a fundamental reorientation of both the principles of product design and the institutional arrangements that govern the delivery of goods and services.

Yet, given the complexity and diversity of modern industrial economies, such a change in outlook is not readily achieved. Government thus has a pivotal role to play in ensuring that such a systems perspective is integrated into its programs of research and education, as well as its regulatory and fiscal policies.

BARRIERS TO CHANGE

Like physical systems, political, social, and economic systems are subject to inertia. Political barriers to change usually mirror broader societal barriers. As an illustration, families of technologies that become integral to the workings of societies tend to remain dominant for many decades. Trains, for example, were the dominant transport system for more than 70 years until they were displaced by trucks and automobiles, which have been dominant for much of this century and are not likely to be displaced soon (Ausubel, 1989). In some cases, once a particular technology path is chosen, the choice may become "locked in," regardless of the advantages of the alternatives (Arthur, 1990).

Technological trajectories are shaped by a variety of economic, social, and political forces. Such trajectories usually cannot be changed without encountering opposition from well-entrenched interests. Reconciliation of these conflicting interests requires the articulation of broad social goals by political leaders, and historically has been achieved only in times of crisis. Thus, harnessing technology in ways that are both productive and ecologically sound is a formidable undertaking, given the inertia or our political and economic structures. However, as society begins to grapple with the potentially serious environmental implications of its economic and industrial practices, the systems perspective provided by industrial ecology is likely to gain a more sympathetic hearing from policymakers.

ENCOURAGING INDUSTRIAL ECOLOGY CHANGES

Practical expression of industrial ecology ideas will have to start at the design level. The product design stage offers a unique point of leverage from which to address environmental problems. Design decisions directly and indirectly determine levels of resource use, types of manufacturing processes, and the composition of waste streams. By giving designers[5] the proper signals about the environmental impacts of their decisions, policymakers can address environmental concerns that arise throughout the product life cycle, from the extraction of raw materials to final disposal.

The principal conceptual questions are when and how policymakers should intervene. Companies already have a number of incentives to move toward "greener" products and processes. For instance, waste-prevention strategies can reduce materials use and energy consumption, and thereby reduce the costs of manufacturing and waste disposal, while limiting potential liability. The consideration of environmental objectives by designers can also have important implications for competitiveness. Market opportunities for environmentally sensitive goods and services are expanding rapidly. Surveys indicate that a substantial percentage of consumers are willing to pay a premium for environmentally sound products.[6]

But despite these incentives for the creation of "green" products and "clean" technologies, a number of technical, behavioral, economic, and informational obstacles need to be addressed. Perhaps the most important obstacles are a variety of market distortions and environmental externalities. For example, government subsidies or preferential tax treatment for the extraction of some virgin materials (e.g., timber and minerals) encourages materials inefficiency. Also, consumers do not pay the full environmental costs of products that are consumed or dissipated during use (e.g., gasoline, agricultural chemicals, and cleaners), nor do they pay the full cost of solid waste disposal. Until such distortions and environmental costs are internalized, environmentally sound design and production decisions are likely to remain economically unattractive.

In general, policymakers can achieve internalization in two ways: by regulation or by economic instruments. Each approach has advantages and disadvantages: regulations, if properly designed, can produce swift and relatively predictable results (e.g., the mandatory phaseout of leaded gasoline), but they can also impose unnecessary costs on industry and stifle environmentally innovative designs. As an example, some environmental regulations discriminate against new technologies by prescribing rigid design standards (the so-called best available technology). Unfortunately, environmental laws have too often been written as if the world were static. Change brings new risks, but the risks to society of not innovating are usually not considered.[7]

Economic instruments, such as surcharges on industrial emissions, can provide flexibility by focusing on desired outcomes rather than methods, but they can be expensive to administer and are often politically unpopular (for example, recall the opposition to the 5 cent gasoline tax, the "nickel for America"). Yet, despite such shortcomings, market-based incentives should be given serious consideration by policymakers. Many environmental problems, such as groundwater contamination or dissipation of heavy metals into the atmosphere, are not caused by large point sources of pollution and therefore cannot be easily addressed by using command-and-control methods. Due to the proliferation of design and materials technology choices, and given that product impacts are almost always multidimensional, policies need to be crafted with flexibility.[8] If particularly acute environmental problems necessitate regulatory action, efforts should be made to design enforcement mechanisms so that innovative and efficient solutions are not ignored. In many cases, market mechanisms can supplement regulatory measures—for example, the excise tax on chlorofluorocarbons. Thus, the challenge for policymakers is to employ a mixture of regulations and economic instruments that encourage designers to take account of rapid technological change while simultaneously safeguarding environmental quality.[9]

THE IMPORTANCE OF A SYSTEMS APPROACH

From an environmental perspective, it is simplistic to view products in isolation from the production and consumption systems in which they function. The greatest environmental gains lie in changing the overall systems in which products are manufactured, used, and disposed of rather than in changing the composition of the products themselves.[10] Product design that accounts for the dynamic relationships among all companies involved in a production system has the potential to produce less waste than product design that takes account of only an individual company's waste stream. This is the central appeal of the industrial ecology concept.

But encouraging a systems approach is a nontrivial undertaking. While in recent years industrial enterprises have demonstrated an increasing ability to manage energy and materials flows in an integrated fashion, these efforts have been highly atomistic, taking place primarily behind the factory gate. In-house waste products are indeed being recycled and reused with greater efficiency and creativity. But there has been little intra-industry and inter-industry coordination in materials management. In some cases there have been regulatory disincentives (e.g., the Resource Conservation and Recovery Act, or RCRA).[11] In other cases, there has been a lack of awareness of the possibilities.

A systems solution to a design problem will often require new patterns of industrial organization, such as the formation of cooperative relationships among manufacturers, suppliers, and waste management providers. The creation of industrial networks can expand the scale of a firm's operations and thereby permit a firm to consider design solutions that would otherwise not be possible. Such cross-company relationships could promote greater materials efficiency in the economy. However, it will not be easy for industry to consider such dramatic changes in its existing production networks. After all, long-standing relationships among manufacturers and suppliers may have to change, and millions of dollars may be invested in the existing infrastructure for production and distribution. Indeed, a systems approach requires a shift in perception by top management such that environmental quality is viewed not as a cost but as a strategic business opportunity.

Government has a key role to play here. First, there is the power of exhortation. Government can encourage new collaborative arrangements across industries and can provide research funds to facilitate such arrangements (e.g., the cooperative agreement among automakers, plastics suppliers, recyclers, and the federal government to explore methods of recovering automotive materials).[12] Next, and probably most important, systems solutions can be encouraged either directly by regulation or indirectly through economic incentives. Recycled content regulations or manufacturer take-back requirements are examples of a regulatory coupling between manufacturing and waste management. The proposal of the German government to require auto manufacturers to take back and recycle their cars, for example, has stimulated the German automakers to rethink the entire industrial ecology of auto production and disposal. As a consequence, new relationships are emerging. Automakers will encourage their material suppliers to accept recovered materials from dismantlers and will specify the use of recovered materials in new car parts, thus "closing the loop." This approach though, may be more appropriate for high-value, durable products with complex material composition than for nondurable or disposable products.[13]

An alternative to take-back regulations involves measures to encourage corporate decision makers indirectly to take a systems approach by using economic instruments to internalize the costs of environmental services (examples include taxes on emissions or virgin materials use; tradable emissions or recycling credits; tax credits; or deposit refund schemes on packaging or hazardous products). This approach would rely on market forces to sort out what new interfirm relationships make sense economically, while giving designers the flexibility to design products with the best combination of cost, performance, and least environmental impact. For example, a substantial carbon tax on fuels could have a dramatic impact on the systems by which products are manufactured, distributed, and disposed of, because fuels are consumed at every stage of the product life cycle.

A FOCUS ON FUNCTION

A true systems view implies a unified consideration of production and consumption activities: supply-side and demand-side requirements need to be treated in an integrated way. This implies a new way of looking at products. The opportunities for linking product design with system-oriented thinking have not been fully explored, but examples are beginning to appear in different sectors of the economy. For instance, pesticide use has declined dramatically where farmers have adopted integrated pest management schemes involving crop rotation and the use of natural predators (U.S. Congress, Office of Technology Assessment, 1990). Due to the success of these new methods, chemical companies are no longer simply supplying pesticides to farmers but are also providing expertise on how to use those chemicals in conjunction with better field design and crop management. In effect, services (i.e., knowledge) have been substituted for chemicals.

Similarly, in the energy supply sector, many utilities are providing energy audit services, and are promoting customer use of energy-efficient equipment, instead of constructing new generating plants. Energy, after all, is used not for its own sake but rather for the services it provides, such as heating, lighting, and transportation. But to encourage decision making on a systemwide basis, utilities need to be allowed to benefit financially from investments in efficient end-use equipment. Recent changes in regulatory frameworks have played a key role in moving utilities in this direction (U.S. Congress, Office of Technology Assessment, 1991b).

The examples of integrated pest management in the chemical sector, and demand-side management in the utility industry, can be applied in a more general way to other industries. When a product is viewed as an agency for providing a service or fulfilling a specific need, the profit incentive changes; income is generated by "optimizing the utilization of goods rather than the production of goods."[14] The notion of thinking about a product in terms of the function it performs is a logical extension of total quality management philosophy. The aim of total quality management is to satisfy customer needs. Customers usually do not care how their needs are met, as long as they are indeed met. Thus, it should not matter whether a customer's requirements are satisfied by a specific product or by a service performed in lieu of that product. (For more on this subject, see Stahel, in this volume.)

Perhaps one of the more intriguing applications of this idea is the "rent model," in which manufacturers retain ownership of products and simply rent them to consumers.[15] By retaining ownership of the products they lease, companies would have a strong incentive to design goods so that they can be reused or remanufactured.[16] However, if the idea of renting rather than selling is to gain currency, public attitudes toward "used" goods will have to change, and government procurement regulations will have to be modified to eliminate biases against refurbished or recycled products. While government-imposed take-back laws would move industry in such a direction, policymakers should proceed cautiously. Although take-back schemes may be a good option for some products, further research on the costs and benefits for a range of products is needed. Mandated design approaches could undermine overall resource efficiency and have unforeseen environmental consequences.

THE CHALLENGE OF GOVERNANCE

A major obstacle to acceptance and application of the principles of industrial ecology is the structure of government itself.[17] Congress, the writer of laws, is organized in a way that works against a systems consideration of environmental problems. Environmental policy is treated in a fragmented fashion. Air, water, and land pollution issues, along with taxes and research, are all under the jurisdiction of separate committees. One consequence of this jurisdictional division is that research needed to implement environmental regulations is often neglected.[18]

This fragmentation extends to the executive branch. The Environmental Protection Agency (EPA) is organized around regulatory responsibilities for protecting air, water, and land; it does not address industries or industrial sectors in a natural way, and its technical expertise in design and manufacturing is slight. The Department of Commerce, on the other hand, is concerned with the competitiveness of industrial sectors, but has little environmental expertise. There is considerable technical expertise in the U.S. Department of Energy's (DOE's) national labs that could be brought to bear on improving design for energy efficicncy and solid waste recycling processes, but environmental quality has not traditionally been a part of DOE's mission.

Recognizing opportunities for systems-oriented design requires that the economic performance and environmental impact of industries or sectors be viewed in an integrated way. Individual companies have little incentive to promote an overall greener vision of their sector. And in general, this cannot be done in the context of a single federal agency. A greener transportation sector, for example, may involve not only improved vehicle fuel efficiency but better management of materials used in automotive, rail, and aviation applications, as well as changes in urban design. An institutional focus above the agency level might spur a more holistic analysis of total sectoral issues, through forums or grant programs.

In Japan the Ministry of International Trade and Industry (MITI), which has responsibility for both trade and competitiveness matters, is also involved in implementing a new recycling law. MITI's involvement is expected to be a strong inducement for companies to comply in a timely way. But in United States, there is no comparable institution that can address trade, competitiveness, and the environment in a coherent fashion. In general, there has been little or no coupling between U.S. technology policy and national environmental requirements.

While it would not make sense to create a separate institution within government to promote industrial ecology concepts, greater coordination between agencies would certainly be desirable. Industrial ecology concepts could be integrated into new interagency initiatives, such as the Manufacturing Technology Initiative and the Advanced Materials and Processing Program announced by the White House in early 1992. In addition, consideration could be given to creating a national environmental technology laboratory (Allenby, unpublished draft).[19]

THE NEED FOR INFORMATION

Government institutions cannot rationally formulate policies without knowing which environmental problems pose the greatest risks. Policymakers currently lack critical information on how materials flow through the economy and about the relative dangers of different materials, products, and waste streams. For instance, there has been considerable concern expressed about the releases of mercury from the incineration of discarded batteries. Yet these releases may be small compared with mercury releases from coal combustion in power plants. Thus, in identifying the major sources of environmental pollutants, it is clear that a systems view is of paramount importance. We need to focus our resources—whether financial or technical—where the risks are greatest, not where the problems are most visible.

CONCLUSION

Since our industrial activities have multiple environmental effects that are not easily disentangled, a systemic approach to environmental policy is needed. But given the barriers enumerated here, such a change in perspective will not come about easily. It is to be hoped that as we increase our understanding of the interconnections between economic and ecological systems, there will be greater political will to develop coherent environmental strategies.

NOTES

1. The world economy is consuming resources and generating wastes at unprecedented rates. In the past 100 years, the world's industrial production increased more than fiftyfold. See Rostow, 1978, pp. 48-49.
2. Industrial ecology refers to the set of relationships among firms in industrial production networks, and the effects of these relationships on the flow of energy and materials through the economy and on the natural world in which the economy is embedded. Some observers envision

systems of industrial production that would emulate the web of interconnections found in the natural world. Nature manages its use of materials in a highly efficient manner. Ideally, production relationships would be organized so that wastes from one process could be used as inputs into other industrial processes.

3. International trade has grown at nearly twice the rate of the gross domestic product over the past decade. About 50 percent of the manufactured imports of the largest industrial countries consist of intermediate, not finished, goods—a reflection of the global nature of production. See OECD (1992).
4. As an example, more than 300 million metric tons of carbon are embodied in the imported manufactured products of six of the largest industrial nations. (Carbon content includes both direct and indirect carbon associated with industrial production.) The sum of carbon embodied in imports for these six countries is about 20 percent of the amount of carbon produced yearly by the United States, surpasses the quantity generated by Japan, and is roughly twice the amount of carbon produced by France. See Wyckoff and Roop (1992).
5. As used here, the term *designers* refers to all decision makers who participate in the early stages of product development. This includes a wide variety of disciplines: industrial designers, engineering designers, manufacturing engineers, and graphic and packaging designers, as well as managers and marketing professionals.
6. See, for example, The Roper Organization, Inc., "The Environment: Public Attitudes and Individual Behavior," a study conducted for S.C. Johnson and Son, Inc., July 1990.
7. The accumulation of knowledge or technological capital can be just as important to future generations as environmental capital. A central ethical question, however, is whether the current generation can fulfill its obligations to future generations by simply substituting technological capital for rapidly disappearing natural capital.
8. There are typically many environmental trade-offs associated with the use of a specific material. For instance, the new classes of high-temperature superconductors, which potentially offer vast improvements in power transmission efficiency and have other promising new applications, are quite toxic; the best of them is based on thallium, a highly toxic heavy metal. The fact that products that use toxic materials can perform socially useful functions, or even have comparative environmental benefits, underscores the need for a flexible approach to environmental questions.
9. For greater detail on the available policy options, see U.S. Congress, Office of Technology Assessment (1992).
10. For example, 80 percent of the waste from a typical fast food chain is produced behind the counter, before food and drinks reach the customer. About 35 percent of the waste generated is corrugated boxes, and another 35 percent is food scrap. Thus, changing delivery methods, and pursuing composting, would have a much greater impact than simply "lightweighting" the packaging of hamburgers. Resources would probably be better spent examining the dynamics of the food chain's distribution and production systems, rather than performing a series of costly life cycle assessments on each of the products used in those systems. Depending on the context, such a systems focus could conceivably result in the elimination of certain products (the ultimate in source reduction), or in the creation of feedback loops that would facilitate recycling and reuse.
11. In the view of some, the Resource Conservation and Recovery Act (RCRA) has impeded the recycling efforts of industry. When a material falls out of a given manufacturing process, it becomes by legal definition a "waste," and is often subject to stiff regulation. The effect of this regulation is to limit any further industrial uses of the material, and, by default, the material really does become a waste.
12. It may be necessary to modify antitrust laws to encourage the formation of research consortia and other collaborative links between industries.
13. Companies such as Xerox and IBM have implemented take-back programs for several years. Because of the high-value, knowledge-intensive nature of their products, these companies have considerable incentive to recover and reuse product subsystems and components. However,

little analysis has been done to determine whether the recovery of low-value items such as food packaging makes environmental or economic sense.

14. Walter Stahel, The Product-Life Institute, Geneva, Switzerland. For more on this idea, see Giarini and Stahel (1989).
15. See note 14 above.
16. Remanufacturing involves the restoration of old products by refurbishing usable parts and introducing new components where necessary. It simultaneously results in product life extension and promotes reuse of subcomponents and materials. Apart from the economic benefits that can accrue to a manufacturer, the reuse of high value-added components takes advantage of the original manufacturing investment in energy and materials. This yields greater environmental benefits than simply recycling the constituent materials of the components.
17. The vitality, adaptability, and resiliency of biological organisms derive not from the mere multiplication of cells but from the efficacy of their organization. This lesson should not be forgotten as we assess the effectiveness of our political "metabolism."
18. For example, volatile organic compounds (VOCs), have been regulated for 20 years, but there has been little actual monitoring of VOC emissions. Emissions are not actually measured but are estimated using models. But the accuracy of these models is in question since the effect of emissions from vegetation is poorly understood. Thus, it is difficult to assess the efficacy of the VOC regulations. See U.S. Congress, Office of Technology Assessment (1989).
19. Several bills have been introduced in the Congress to create such an agency.

REFERENCES

Allenby, Braden. Unpublished draft. Why We Need a National Environmental Technology Laboratory (And How to Make One).

Arthur, W. Brian. 1990. Positive feedbacks in the economy. Scientific American 262(2):92–99.

Ausubel, Jesse. 1989. Regularities in technological development: An environmental view, Pp. 70-91 in Technology and Environment, Jesse Ausubel and Hedy Sladovich, eds., Washington, D.C.: National Academy Press.

Ayres, Robert U. 1992. Toxic heavy metals: Materials cycle optimization. Proceedings of the National Academy of Sciences 89(3):815-820.

Frosch, Robert A. 1992. Industrial ecology: A philosophical introduction. Proceedings of the National Academy of Sciences 89(3):800–803.

Galloway, James N., J. David Thornton, Stephen A. Norton, Herbert L. Volchok, and Ronald A. N. McLean. 1982. Atmospheric Environment 16(7):1678.

Giarini, Orio, and W. Stahel. 1989. The Limits to Certainty: Facing Risks in the New Service Economy. Boston, Mass.: Kluwer Academic Publishers.

Organization for Economic Cooperation and Development. 1992. The International Sourcing of Intermediate Inputs. DSTI/STII/IND(92)1, Paris (January).

Ponting, Clive. 1990. Historical perspective on sustainable development. Environment 32(9).

Rostow, W. W. 1978. The World Economy: History and Prospects. Austin, Texas: University of Texas Press.

U.S. Congress, Office of Technology Assessment. 1989. Catching Our Breath: Next Steps for Reducing Urban Ozone. Washington, D.C.: U.S. Government Printing Office.

U.S. Congress, Office of Technology Assessment. 1990. Beneath the Bottom Line:

Agricultural Approaches to Reduce Agrichemical Contamination of Groundwater. OTA-F-418. Washington, D.C.: U.S. Government Printing Office.

U.S. Congress, Office of Technology Assessment. 1991a. Changing by Degrees: Steps to Reduce Greenhouse Gases. OTA-0-482. Washington, D.C.: U.S. Government Printing Office.

U.S. Congress, Office of Technology Assessment. 1991b. Energy Technology Choices: Shaping Our Future. OTA-E-493. Washington, D.C.: U.S. Government Printing Office.

U.S. Congress, Office of Technology Assessment. 1992. Green Products by Design: Choices for a Cleaner Environment. OTA-E-541. Washington, D.C.: U.S. Government Printing Office.

Wyckoff, Andrew W., and J. M. Roop. 1992. The Embodiment of Carbon in Imports of Manufactured Products: Implications for International Agreements on Greenhouse Gas Emissions. OECD, Paris (May).

Emerging Industrial Environmental Practice

The Greening of Industrial Ecosystems. 1994.
Pp. 137–148. Washington, DC:
National Academy Press.

Integrating Environment and Technology: Design for Environment

BRADEN R. ALLENBY

Twenty years ago, environmental problems appeared simple and obvious: the Potomac and Hudson Rivers were so polluted that they could not be fished; the air in Los Angeles was foul; Love Canal and other sites were poisoned by toxic chemicals. Remedies were equally direct: the Clean Water Act, the Clean Air Act, Superfund, and the Resource Conservation and Recovery Act.

Now, with several decades of research and experience behind us, we have begun to recognize that we are treating the symptoms, not the disease. The problem is not individual rivers, airsheds, or hazardous waste sites, although they must be addressed. The problem is the relationship between human economic activity and the environment. It is a systems problem, arising from fundamental changes in the scale of human activity in relationship to supporting biological, physical, and chemical systems. Regional and global environmental perturbations cannot be adequately mitigated until this relationship is understood and regulatory and industrial practices are modified to reflect such an understanding.

A basis for developing a broad understanding of these issues, frequently termed "industrial ecology," is being established, albeit the effort is still in its infancy (Allenby, 1992a; Ayres, 1989; Frosch and Gallopoulos, 1989). Nonetheless, five fundamental principles upon which we may base development of improved methodologies for integrating technological and environmental systems can already be identified:

1. Methodologies should be comprehensive and systems-based.
2. Methodologies should be multidisciplinary, including technical, legal, economic, political, and cultural dimensions to the extent possible.

3. Mitigation of environmental perturbations can only be achieved by focusing on technology, and developing policies and practices that encourage the evolution of environmentally preferable process and product technologies.

4. Economic actors, including private firms, must internalize environmental considerations and constraints to the extent possible, given existing exogenous constraints on firm behavior (e.g., laws such as the antitrust statutes or the prices of inputs and competitive products).

5. Policies and regulations must reflect the need for experimentation and research as different paths and methodologies are tried (rigid micromanagement through command-and-control regulation will in many cases be incompatible with systems-based approaches embodied in methodologies such as Design for Environment (DFE), as the unfortunate negative impact of the Resource Conservation and Recovery Act on industrial recycling practices demonstrates (see the paper by Pfahl in this volume and, more generally, Office of Technology Assessment [OTA], 1992).

Obviously, a great deal of work remains to be done before we begin to fully understand industrial ecology and apply it to current industrial ecosystems (see the discussion by Ehrenfeld in this volume). In the interim, however, there can be no excuse for evading the application of the approaches we do understand to ongoing regulatory and industrial behavior. Although just emerging and still generally untested at this point, DFE is one potential methodology for accomplishing this. Design for Environment should be regarded more as an approach, than as an existing, implemented system.

Before introducing DFE, it is important to recognize that our current unsophisticated approaches tend not to recognize important differences among classes of products and materials. Thus, for example, it is useful to differentiate between two classes of manufactured items: low-design/high-material items, such as packaging, consumer personal care items (e.g., soaps and shampoos), and bulk chemicals; and high-design/low-material products such as automobiles, electronic and communications equipment, and airplanes.

These product streams generally have different life cycles within the economy. For example, the former tend to be used up and dispersed into the environment rather than discarded (packaging being an obvious exception). Materials use in the two cases also implies different recycling and disposal requirements. Materials use in a low-design/high-material product stream tends to be relatively simple, and, in many cases, few materials are incorporated into individual items (this need not be the case, however; a snack chip bag only 0.002 inch thick consists of nine separate layers of material [OTA, 1992]). Thus, recycling is fairly straightforward. On the other hand, the structure and materials use of a high-design/low-material product such as a printed wiring board are highly complex, and many materials, including different plastics, ceramics, alloys, frits, and glasses, are often present in a single product. Recycling of materials for such a product is thus much more complicated and difficult.

Moreover, as implied by the designation, the design function is far less important for a low-design/high-material product than for a high-design/low-material product. This reflects a fundamental difference: the design of a complex high-design/low material product requires consideration of an additional, highly complex, system that low design/high-material products do not—that is, the product itself. Associated upstream and downstream product and manufacturing process implications of changes to such a product or its associated manufacturing systems make the analysis far more complex than in the case of a low-design/high-material item. Moreover, the supplier/customer networks for high-design/low-material products tend to be far more complex than those for low-design/high-material products.

This has a number of policy implications, the primary one being that regulatory tools that may be effective in encouraging environmentally appropriate technology for high-design/low-material items may be dysfunctional or at least inefficient when applied to many low-design/high-material products. Thus, for example, post-consumer take-back requirements for high-design/low material articles will drive firms (and thus designers) to produce products more easily disassembled and recycled. On the other hand, postconsumer take-back requirements imposed on low-design/high-material products such as consumer packaging, which generally require significant transportation of used product, may not be economically efficient.

DESIGN FOR ENVIRONMENT

As a practical matter, the only way the five principles enumerated above can be implemented by industry in practice, at least for high-design/low-material products and associated manufacturing processes, is by driving environmental considerations and constraints into the design process. Implementation of DFE practices is intended to accomplish this.

The idea behind DFE is to ensure that all relevant and ascertainable environmental considerations and constraints are integrated into a firm's product realization (design) process. The goal is to achieve environmentally preferable manufacturing processes and products while maintaining desirable product price/performance characteristics (Allenby, 1991).

DFE is intended as a module of an existing design system known as "Design for X," or DFX, where "X" is a desirable product characteristic such as testability, manufacturability, or safety (Gatenby and Foo, 1990). This has at least two critical advantages. For one, linking implementation of DFE to an existing process reduces the culture shock of integrating environmental considerations into product and process design and makes the entire package far more acceptable and easier to implement. For another, the need to create DFE to fit into an existing design procedure imposes a necessary discipline on DFE development that helps make DFE practical in a real operating environment.

Implementing DFE has a number of benefits for the firm. The major benefit is that it provides a mechanism for the firm to manage environmental issues as they evolve from simply overhead, "end-of-pipe," considerations to strategic and competitively critical. Thus, compliance with increasingly complex, sometimes contradictory, environmental regimes can be eased by designing to avoid emissions. Costs arising from taxes, fees, or burdensome regulations directed at inputs such as energy or virgin materials, or from wastes or carbon produced during manufacturing, can be identified and managed only through such a comprehensive approach.

Another important reason firms are adopting DFE is the need to meet increasing environmental concerns on the part of customers. There are two major components to this challenge for manufacturing firms.

The first is the much-discussed growing interest that consumers exhibit in "green" products. At this point, determining environmental preferability, or "greenness," over the life cycle of a product is frequently beyond the state of the art. Moreover, determining the depth and importance of environmental consumerism is somewhat problematic and depends to a great extent on the product and its market, as well as overall economic conditions. It is, however, clear that such preferences exist to some degree.

The second is more subtle and much harder to address ad hoc: the demands of sophisticated customers for products that reduce their potential environmental costs. Thus, for example, the U.S. Air Force may be concerned about a weapons system that requires chlorofluorocarbons (CFCs) for routine maintenance, or that produces significant used oil or chlorinated solvent waste streams. Sophisticated business customers may pressure component or subassembly suppliers to reduce their use of toxic substances such as lead solders or batteries containing cadmium, or to use environmentally preferable packaging. State and national laws require recycling of batteries and elimination of heavy metals in packaging or plastics. Such requirements frequently require design changes: they clearly cannot be addressed by traditional "end-of-pipe" means.

Comparing DFE with other environmental management and regulatory concepts helps demonstrate the comprehensive nature of the methodology. Perhaps the most familiar of these concepts is "pollution prevention" and such associated terms as "waste minimization" and "toxics use reduction." DFE incorporates these concepts, but as elements in a more complex multidimensional analysis. To a large extent, these concepts still reflect the "end-of-pipe" mind-set, in that they do not comprehend the need for a comprehensive, systems-based, technologically sophisticated approach to environmental management in a high-technology economy.

Other concepts involve aspects of the comprehensive DFE system as well. These include, for example, practices such as "Design for Disassembly," "Design for Refurbishment," "Design for Component Recyclability," and "Design for Materials Recyclability." Like pollution prevention, these are obviously necessary dimensions of a DFE system.

These single dimensions are important not just in themselves but because they offer firms a relatively easy path to begin implementing DFE. No firm has yet implemented a comprehensive DFE system, and, indeed, fully implementing DFE practices will in all likelihood require that most firms develop new competencies, organizations, and information systems. This will take time and, in many cases, changes in organizational cultures as well. Accordingly, the most practical path for firms to follow may be to concentrate on implementating a few of these concepts initially, such as pollution prevention and Design for Disassembly, and then move on to other aspects of DFE as experience is gained.

Life cycle analysis, or LCA, incorporates into DFE the concept that all environmental impacts of an item—from those attributable to inputs, to manufacture, to consumer use, to disposal—should be considered in evaluating the environmental preferability of the product (Society of Environmental Toxicology and Chemistry [SETAC], 1991). LCA is a means by which data on environmental impacts, an important component of a DFE approach, can be generated. LCA can thus be equated to the DFE Matrix System discussed below. While it is probable that both LCA and the DFE Matrix System will continue to evolve in similar directions, at present LCA appears to apply more to low-design/high-material products, in that proposed methodologies do not reflect the systems implications of manufacturing complex articles. Perhaps for similar reasons, LCA also is essentially an analytical method; it does not provide a mechanism by which identified environmental considerations and constraints can be introduced into product and process design, as DFE does.

A number of other generic terms, such as "green engineering" and "environmentally conscious manufacturing," are also in use. Although these efforts have generated useful ideas and specific technical practices, as currently defined they tend to follow a rather ad hoc approach to meeting environmental goals. The lack of a systems-based approach makes it difficult for firms, and designers, to implement these concepts in a comprehensive manner.

IMPLEMENTING DESIGN FOR ENVIRONMENT

The implementation of DFE practices involves two categories of activities: (a) global, comprehensive projects whose effect extends across all design functions; and (b) specific individual evaluations of products, processes, or inputs.

In the first category are projects that can be undertaken at any time and will help result in environmentally preferable products across the board. Such activities include review of all internal specification documents to determine whether unnecessary environmentally harmful processes (such as cleaning with chlorinated solvents) or components (lead solder where conductive epoxy systems might be used) are being required. Similarly, manufacturing firms should evaluate specifications and requests for proposals from their customers to determine whether unnecessary environmental impacts are being explicitly or implicitly required.

Using generic contract clauses to change supplier behavior can also be effective. For example, several years ago AT&T began by contract to require that all suppliers use non-CFC packaging. Those who were unable to comply contacted AT&T, which then directed them to preferable alternatives. A general improvement of all packaging used by both AT&T and its suppliers was achieved.

Identifying unnecessary process steps can also be an effective way of reducing environmental impact across many operations. Thus, many electronics companies found they could reduce their use and emissions of CFCs and chlorinated solvents by eliminating some cleaning steps or by doubling-up cleaning operations. This activity can be driven back into the design process by identifying design decisions that require environmentally harmful processing activities, and selecting alternatives. Thus, for example, use of an open relay switch on a printed wiring board requires that the board be cleaned with a chlorinated solvent, since such relays "can't swim"; that is, they cannot be exposed to water. Substituting a sealed relay permits the use of environmentally preferable aqueous cleaning systems.

More specific DFE activities involve an analysis of options for specific design choices. Here, a particular product, process, or input would be evaluated using several basic steps.

1. *Scoping*. The target product, process, or input is chosen, and options are identified. This is a critical step, particularly because the process of identifying options also generates a set of potential competitive surprises.

A second important component of the scoping process is to determine the depth of analysis required. For a fundamental design decision—such as moving away from lead solder technologies in printed wiring board assembly—a fairly rigorous analysis would be appropriate. However, where incremental changes to an existing product or process are under consideration, there are relatively few options open to the designer, and a DFE analysis can be correspondingly limited.

2. *Data Gathering*. The next step is to gather and evaluate all relevant data. This may be done using LCA methodologies or, alternatively, by completing a data collection effort aimed specifically at DFE information needs. Such an effort revolves around collecting information in at least four areas, illustrated by the data collection matrices in Figures 1–5: environmental primary, manufacturing primary, social/political primary, and toxicity/exposure primary. A summary matrix then captures the most severe concern for each ranking. The data in the matrices shown in Figures 1–5 are drawn from an example developed by the author and described later in this chapter.

There are three possible entries for each matrix cell (Figure 6). One or two pluses in a cell means the option has positive environmental effects for that cate-

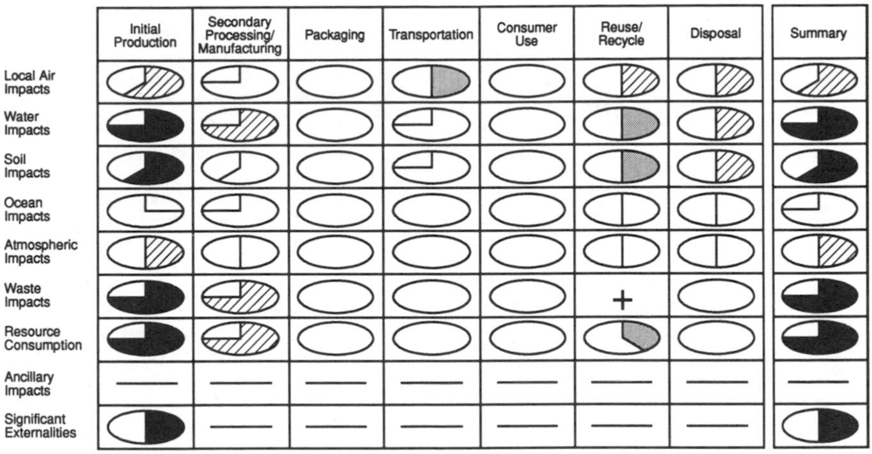

	Initial Production	Secondary Processing/ Manufacturing	Packaging	Transportation	Consumer Use	Reuse/ Recycle	Disposal	Summary
Local Air Impacts								
Water Impacts								
Soil Impacts								
Ocean Impacts								
Atmospheric Impacts								
Waste Impacts						+		
Resource Consumption								
Ancillary Impacts	—	—	—	—	—	—	—	—
Significant Externalities		—	—	—	—	—	—	

FIGURE 1 Environmental primary matrix for bismuth.

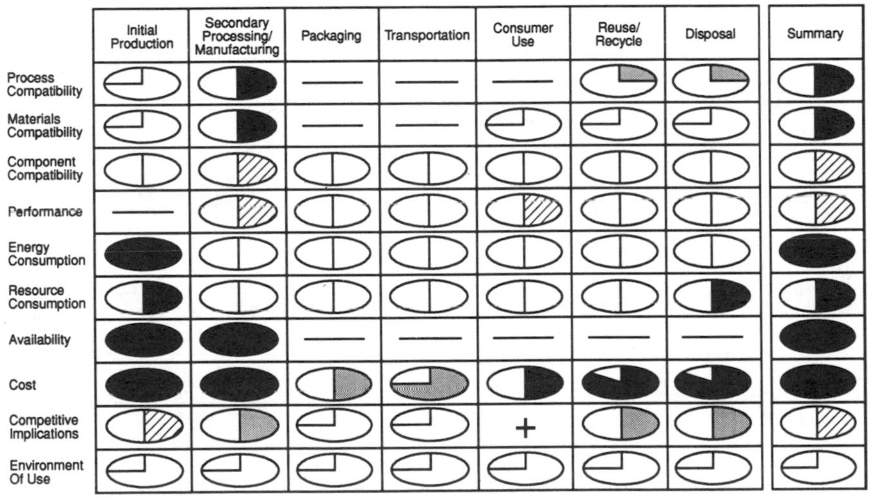

	Initial Production	Secondary Processing/ Manufacturing	Packaging	Transportation	Consumer Use	Reuse/ Recycle	Disposal	Summary
Process Compatibility			—	—	—			
Materials Compatibility			—	—				
Component Compatibility								
Performance	—							
Energy Consumption								
Resource Consumption								
Availability			—	—	—	—	—	
Cost								
Competitive Implications					+			
Environment Of Use								

FIGURE 2 Manufacturing primary matrix for indium.

gory and indicates the relative degree of benefit. A straight line means the cell is inapplicable to the option. An oval indicates some degree of concern: an open, or blank, oval indicates minimal concern; dots indicate some concern; diagonal lines indicate moderate concern; and solid black indicates serious concern.

An important feature of this graphical approach is the ability to indicate relative degrees of uncertainty. Thus, the amount by which each oval is filled in

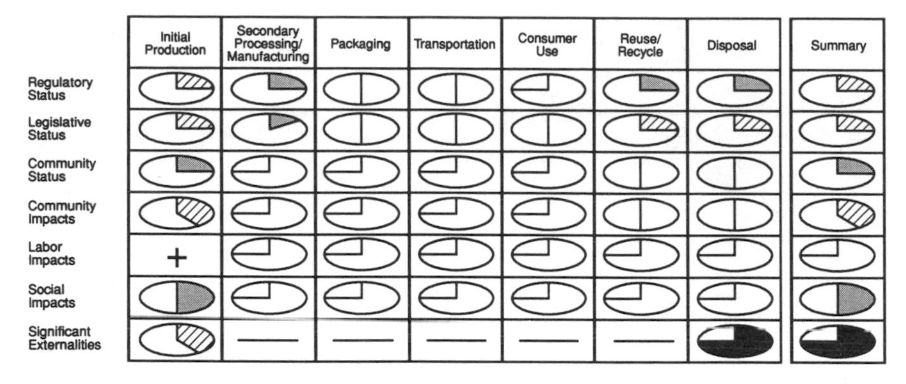

FIGURE 3 Social/political primary matrix for indium.

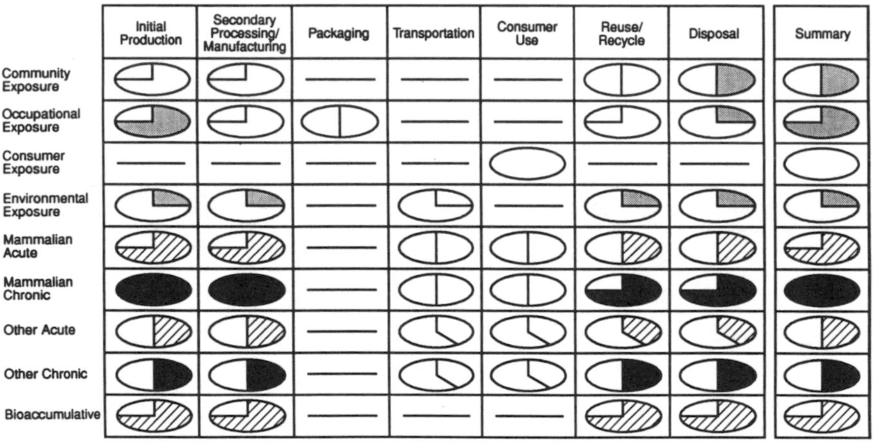

FIGURE 4 Toxicity/exposure primary matrix for lead.

(going clockwise from 12 o'clock, as illustrated in Figure 6) with the appropriate pattern indicates the relative degree of uncertainty. For example, a solvent with potentially serious health effects, such as methylene chloride (a suspected carcinogen), would receive a solid black rating in the appropriate cell but, as those effects are quite uncertain, the oval might be only be half filled.

3. *Data Translation.* Once an analysis is completed, the information must be digested and turned into tools with which the design engineer or team can work. It is not very helpful, for example, to tell a designer that methylene chloride is a possible human carcinogen if the designer is unaware of alternatives. It is far more effective to create design tools, such as standardized components lists; design procedures, such as checklists that can be reviewed by environmental ex-

	Bismuth	Indium	Lead	Epoxy
Toxicity / Exposure				
Environmental				
Manufacturing				
Social / Political				

FIGURE 5 Summary matrix of the analysis.

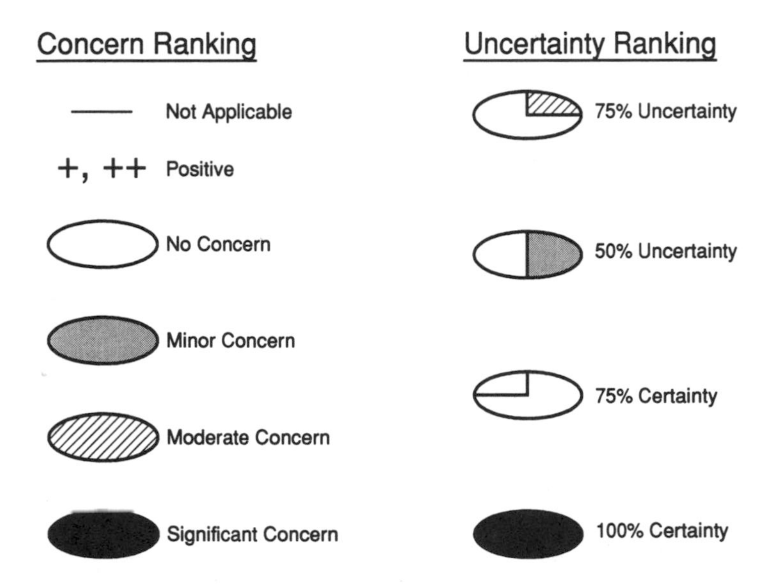

FIGURE 6 DFE Information System (DFEIS) symbols.

perts; and software systems that incorporate these data into the design process without requiring further knowledge on the part of the designer.

These design tools can take many forms. For example, as AT&T phased out its use of CFCs in manufacturing, it created two useful tools. One, a handbook, listed all common solvents and, among other things, provided a color flag to indicate whether use of the material was acceptable from an environmental and regulatory perspective (for example, benzene would be red, or not acceptable, while water would be green, or acceptable under all circumstances). The other was a heuristic hierarchy of process choices (implementation of which frequently entailed changes in design process or product design). The preferable choice was to eliminate the cleaning step entirely; the second was to use an aqueous system; the third was to use a semiaqueous system (water/hydrocarbon); the fourth, temporary and least preferred, was to use a chlorinated solvent.

Another "design tool" is simply to require that all components be labeled to ease postconsumer recycling. Thus, for example, plastic casings for computers or copiers increasingly have a label embossed on the inside giving the type of plastic, which facilitates material recycling. This need not concern the designer directly; it merely needs to be incorporated into the standard components list or required of the supplier through specifications.

DESIGN FOR ENVIRONMENT: TESTING THE SYSTEM

An experimental test of a formal DFE Information System (DFEIS) methodology was carried out by the author on a fairly controversial problem that AT&T confronted regarding the desirability of substituting indium alloys, bismuth alloys, or isotropic conductive epoxy technologies (using silver as the filler) for the lead solder currently used in assembly of printed wiring boards in electronics manufacturing (Allenby, 1992b). The results, presented in Figure 5, were counterintuitive, supporting a conclusion that significant substitution of indium, bismuth, or isotropic conductive epoxy technologies for lead solder would not be environmentally preferable.

None of the environmental effects that led to this conclusion occurred during the manufacturing or consumer use life cycle stages, the only ones that would have been considered in most standard industrial evaluations. Rather, they reflected basic characteristics of indium and bismuth ores (and, to a lesser degree, silver); namely, the low concentrations of indium and bismuth in virtually all ores, and the low world reserves of the metals taken as a whole. Under these circumstances, extraction would be very energy-intensive, and substantial environmental impacts, at least locally, could be anticipated (see Figure 1). Moreover, this raises obvious cost and availability concerns that should be addressed before relying on these options for a critical manufacturing function (see Figure 2).

Additionally, it was noteworthy that even this relatively limited, real world test of the DFEIS raised several unresolved ethical issues. For example, if indium or bismuth were to be substituted for lead solder, much of the environmental impact (and associated social and community impacts as well) would fall on localities where the mining and processing occurred, including some outside the United States. The potential environmental benefits, a reduction in the amount of lead solder from printed wiring boards in waste streams, would accrue primarily to localities near landfills or incinerators in the United States, where the electronic items would be disposed of. The asymmetrical geographic distribution of risk and benefit raises clear equity concerns, but I am aware of no generally accepted methods for resolving them.

A second issue, that of intergenerational resource allocation, is an extremely difficult equity problem. Since there is little indium in the world, is it appropriate to consume much of this limited stock in the manufacture of televisions and VCRs, where it will eventually end up in sinks from which it is virtually unrecoverable? Is this acceptable given that the risks posed by alternatives, such as lead solder, are probably minimal for this use? In essence, this amounts to buying little or no risk reduction for U.S. inhabitants today by loading potentially substantial costs on future generations. If other materials can substitute for indium completely, the answer might still be yes, but we have no way of knowing this: indium might have some unique value in the future. (This is the rationale behind the ranking of "serious concern" in the disposal life cycle stage/significant externalities category in Figure 3.)

At this point, at least, there are no generally accepted legal or economic answers to this conundrum. What is perhaps more interesting, however, is that this very difficult issue arose in a relatively simple analysis of a serious existing design decision. Companies are deciding now whether to begin the difficult task of shifting to alternative joining technologies for printed wiring boards, so they will implicitly answer the ethical questions in any event. By failing to implement DFE practices, we are not avoiding problems; we are just ignoring them.

The toxicity/exposure primary matrix, containing the most familiar material, completes the matrix set, but even here there were important implications. Thus, as Figure 4 indicates, even for lead, a material whose toxic effects have been known since Roman times, considerable uncertainty over nonmammalian toxicity remains. Also it should be noted that data on exposure and on acute toxicity are integrated to provide the rankings captured in the summary matrix. This reflects the well-known toxicological principle that the hazard posed by any agent is a function of both the inherent toxicity of the agent and the concomitant exposure of the target population. Either exposure or toxicity alone does not generate hazard.

CONCLUSION

We are now at a stage where we are beginning to understand the multidisciplinary, systemic nature of the interrelationship between technology and environment. Current regulatory and industrial practices, however, are still predicated on an increasingly dysfunctional symptoms-oriented approach. Design for Environment practices provide a means by which industry—and regulators—can move beyond current single-dimensional, ad hoc approaches in the short term and evolve a systems-based, comprehensive set of practices over the long term. Both the environment and industrial competitiveness will benefit as a result.

REFERENCES

Allenby, B. R. 1991. Design for environment: A tool whose time has come. SSA Journal September:5–9.

Allenby, B. R. 1992a. Industrial ecology: The materials scientist in an environmentally constrained world. MRS Bulletin 17(3):46–51.

Allenby, B. R. 1992b. Design for Environment: Implementing Industrial Ecology. Ph.D. dissertation, Rutgers University.

Ayers, R. U. 1989. Industrial metabolism. Pp. 23–49 in Technology and Environment, J. H. Ausubel and H. E. Sladovich, eds. Washington, D.C.: National Academy Press.

Frosch, R. A., and N. E. Gallopoulos. 1989. Strategies for manufacturing. Scientific American 261(3):144–152.

Gatenby, D. A., and G. Foo. 1990. Design for X: Key to competitive, profitable markets. AT&T Technical Journal 63(3):2–13.

Office of Technology Assessment. 1992. Green Products by Design: Choices for a Cleaner Environment. Washington, D.C.: U.S. Government Printing Office.

Society of Environmental Toxicology and Chemistry. 1991. A Technical Framework for Life-Cycle Assessment. Washington, D.C.: SETAC Foundation.

The Greening of Industrial Ecosystems. 1994.
Pp. 149–164. Washington, DC:
National Academy Press.

Preventing Pollution and Seeking Environmentally Preferable Alternatives in the U.S. Air Force

EDWARD T. MOREHOUSE, JR.

To seek environmentally preferable alternatives to its current business practices, the U.S. Air Force implemented a comprehensive Pollution Prevention Program. The program draws together environmental pronouncements embodied in executive orders, legislation, and Department of Defense (DOD) and Air Force policy statements, and it overlays sound business principles to produce a comprehensive program designed to integrate environmental considerations into the fabric of Air Force business. The Air Force model may be broadly applicable to other large government entities, and to private firms as well.

A number of environmental issues surfaced in approximately the same time frame, which provided the Air Force Pollution Prevention Division with an opportunity to propose a comprehensive program and successfully sell it to senior Air Force leadership. The most compelling issue was the renegotiation of the Montreal Protocol on Substances That Deplete the Ozone Layer, which calls for a complete production phaseout of ozone-depleting chemicals (ODCs). Since virtually every military system is dependent to some degree on ODCs, this issue became the first priority of the Pollution Prevention Program. Other issues included data indicating the unit cost of hazardous waste disposal had increased tenfold over a five-year period, the emergence of the Environmental Protection Agency (EPA) voluntary toxics reduction (33/50) program, passage of the Pollution Prevention Act, dramatic increases in costs of cleaning up contaminated waste sites, and the rapid disappearance of space to dump municipal solid waste. These issues taken together pointed to the need for a comprehensive, Air Force-wide program aimed at all environmental pollutants and supported at the highest levels of Air Force leadership.

By engaging the entire Air Force organization, the program encompasses all of the various roles played by a large federal agency, which include researcher, technology and product developer, large consumer of commercial and industrial products and technologies, operator of large industrial plants, construction agent, facility manager, and city manager. In each of these roles, the Air Force not only affects the environment but also has an opportunity to lead others, both inside and outside government, by example.

This paper describes the Air Force Pollution Prevention Program, explains how it was developed and is being implemented, and provides a perspective for assessing its significance.

DOD PROGRAM POLICY AND OBJECTIVES

On October 10, 1989, Secretary of Defense Dick Cheney issued a memorandum on environmental management policy to the secretaries of the military departments. The policy statement was direct and clear. Secretary Cheney said the "Administration wants the United States to be the world leader in addressing environmental problems, and I want the Department of Defense to be the federal leader in agency environmental compliance and protection." The statement made the point that the Department of Defense, as the largest federal agency, has a great responsibility to meet this challenge and that Secretary Cheney wanted every command to set an environmental standard by which federal agencies would be judged. One of the most significant statements in the policy is "the first priority of our environmental policy must be to integrate and budget environmental considerations into our activities and operations." This provided a policy basis for identifying environmental investments needed to meet environmental goals and obtaining needed financial resources.

Complementing Secretary Cheney's policy is the DOD Directive 4210.15, Hazardous Materials Pollution Prevention. It states that the Military Services are to "select, use and manage hazardous materials so as to incur the lowest life cycle cost to protect human health, the environment and long-term liability." While this is a basic and sound business approach, implementation is an enormously difficult and complex task. Since the art and science of selecting materials and processes to minimize environmental impact are in their infancy, there was no implementing guidance to accompany the directive. This left significant discretion to program managers in a world of many competing priorities.

The activity that drives the greatest amount of hazardous waste generation in the U.S. military is system acquisition. The technologies used to develop, build, operate, and maintain weapons systems drive approximately 90 percent of the military use of hazardous materials/use and generation of hazardous waste. At first, the notion of a "green" weapons system may seem absurd, but in reality it is not. Weapons systems spend most of their lives in a peacetime role. Consequently, the environmental impact created by the system results in large part from training and maintenance. Many systems, aircraft in particular, often remain in the inventory for 30 years or more. During that time, maintenance and refurbishment performed at Air Force industrial plants use large quantities of hazardous materials and generate large quantities of waste. The material and process selections made during system development dictate maintenance and operations procedures and define the environmental impact for the entire 30 years of ownership. These material decisions are made by program offices responsible for delivering new systems that meet performance criteria, on time, and within budget. DOD Instruction 5000.2, which prescribes acquisition procedures, includes a requirement to include hazardous material considerations in acquisition activities. As with the Hazardous Materials Pollution Prevention Directive, it is not specific about how to accomplish the task.

LAUNCHING A NEW AIR FORCE POLICY

On November 13, 1992, the chief of staff of the Air Force, General Merrill A. McPeak, and the secretary of the Air Force, Donald B. Rice, jointly signed a memorandum to the commanders of all major commands, all assistant secretaries, and all deputy chiefs of staff proposing a comprehensive Pollution Prevention Action Plan. The memorandum asked for endorsement of specific goals and objectives, recommendations to improve the proposed program, and an estimate of the resources needed to achieve the goals. The Pollution Prevention Division used these endorsements and resource requirements to establish an investment strategy and to convince the Secretary of the Air Force that the projected economic and environmental benefits warranted the investment. On January 7, 1993, the chief of staff and the secretary of the Air Force signed a memorandum formally establishing the program. In the memorandum, they stated, "The Air Force is committed to preventing future pollution by reducing use of hazardous materials and releases of pollutants into the environment to as near zero as feasible. We must mobilize our whole team and find ways to move faster."

HIGHLIGHTS OF THE AIR FORCE POLLUTION PREVENTION PROGRAM

There are six objectives to the Pollution Prevention Action Plan. The first is to reduce the use of hazardous materials in all phases of new weapons systems from concept through production, deployment, and ultimate disposal. It includes subobjectives describing specific actions and milestones that lead to institutionalizing pollution prevention in the systems development and acquisition process. The assistant secretary for acquisition is responsible for achieving this objective and has authority to make the necessary changes in acquisition policy and procedures.

The second objective is to "reduce the use of hazardous materials in existing weapons systems by finding less hazardous materials and processes and integrating them into Technical Orders, Military Specifications and Standards." (Technical Orders are like owners manuals. They prescribe maintenance processes, including material prescriptions.) Most existing weapons systems were developed, built, and acquired before environmental considerations were part of the material decision process. Under this objective, their maintenance and operations procedures will be revisited and materials and processes changed to environmentally preferable alternatives. As with the first objective, there are many subobjectives with milestones. Responsibility and authority belong to the deputy chief of staff for logistics.

The third objective deals with the Air Force role as operator of installations, and requires the reduction of "hazardous material use and waste generation at installations and government-owned, contractor-operated (GOCO) plants." Subobjectives were negotiated with field commanders and responsible functional area leaders and prescribe reductions in volatile air emissions (50 percent by 1999), municipal solid waste disposal (25 percent by 1996 and 50 percent by 1997), purchase of ODCs (eliminate by June 1993 or April 1994 depending on end use), purchases of the 17 identified industrial toxics of the EPA 33/50 program (50 percent reduction by 1996), and generation of hazardous waste (25 percent reduction by 1996 and 50 percent by 1999). This objective also establishes a requirement to characterize waste streams to all media, so the Air Force can quickly address wastes that the EPA may target for future regulation, voluntary reduction programs, or reporting requirements. A comprehensive characterization of waste streams also provides data needed to identify future pollution prevention opportunities as the program evolves. Responsibilities to meet these reduction goals apply to each of the three major organizations in the Air Force that generate the pollution. Achieving the target for the 17 toxics in the 33/50 program at GOCOs should be easy since virtually all contractors who operate Air Force plants have voluntarily signed onto the program.

The fourth and fifth objectives say that the best pollution prevention technologies either be acquired from the private sector, other government agencies, or other nations or be developed internally and shared with others. Subobjectives challenge users and maintainers of weapons systems, support systems, and infrastructure to identify technologies needed to enable them to achieve the program goals and make those known to the research and development community. Again, responsibility is assigned according to lines of authority.

The final objective is to "establish an Air Force investment strategy to fund the Pollution Prevention Program." The Air Force Engineer is responsible for overall implementation of the Pollution Prevention Program and for working with the Acquisition and Logistics communities to ensure the financial requirements are identified and programmed into the budget.

The key to translating policy into practice is to change behavior, both within the Air Force and among the suppliers. For this program to be successful, all Air

Force employees need to understand the environmental requirements of their job and how to comply; all family members living in military housing need to understand how their behavior affects the environment and how to change it appropriately; and all contractors doing business with the Air Force need to understand how the environmental performance of their companies and their products will be factored into Air Force purchasing decisions. Each of these behavioral changes represents a major education and training challenge.

POLLUTION PREVENTION MEANS CULTURAL CHANGE

The Air Force is using several techniques for changing internal behavior. They include education and training, changes to accounting systems that "make polluters pay" and give "corporate" visibility to environmental compliance costs, creating pollution prevention scorecards that measure organizational progress toward meeting pollution prevention objectives, and using environmental protection committees (EPCs) to give leadership and high level visibility to how well organizations are meeting the program objectives.

Education and Training

Most people care about the environment and want to do the right thing environmentally. Showing people how to do their jobs in a way that protects the environment is one of the cornerstones of the program. Accordingly, the internal environmental education and training program reaches each and every Air Force employee. Different jobs have different requirements, and the curricula and methods for delivering the training are tailored accordingly. All recruits entering the Air Force receive environmental training as part of their basic military training. Airmen and officers whose jobs involve dealing with hazardous materials receive more in-depth, specialized training. The Air Force Institute of Technology (AFIT) offers a variety of environmental short courses, including one specifically on pollution prevention. Each installation pollution prevention manager is encouraged to take this course, and central funding is provided as an incentive. AFIT also offers master of science degrees in environmental management. Traveling teams from the Air Force Center for Environmental Excellence conduct regional training courses, in cooperation with the EPA, for all individuals at installations responsible for executing pollution prevention programs at the installation level. The Air Force Academy also offers degrees in environmental science and environmental management. The Reserve Officers Training Corps offers scholarships to students studying environmental sciences.

Courses required for enlisted members to advance in rank include environmental training. The Air Training Command conducts 105 technical training courses specifically for enlisted members and provides detailed training on hazardous material and hazardous waste management. Orientation courses required for people transferring from one installation to another include a session on the environmental program at their new base. The Defense Systems Management Course (DSMC), which provides basic education required for all acquisition officers, includes a section on the environmental impact of acquisition programs. However, it must be expanded to emphasize pollution prevention opportunities and to show how to incorporate pollution prevention into acquisition programs.

In addition to educating the workers, the Air Force has focused on educating leadership at all levels through a unique course called the Environmental Leadership Course, conducted by a general officer. Between 1990 and 1992, 15 courses were conducted for commanders and their staffs, and attendance exceeded 800. With only nine major commands and approximately 100 installations, coverage was very good. Each course included a presentation on the Pollution Prevention Program to explain why it is important to commanders, the financial benefits of a successful program, and tips for commanders to use to assess the health of their programs. This one- to two-day course had a significant effect on the attention commanders give to their pollution prevention programs.

The Environmental Comprehensive Assessment and Management Program (ECAMP) and opportunity assessments are additional tools for educating installations about their pollution prevention opportunities. Installations conduct an ECAMP self-assessment once a year using internal staff, and once every three years using external independent assessors. The assessment team uses a series of protocols to evaluate every aspect of the installation environmental program. The results of the ECAMP are for commanders to use to improve their programs. Opportunity assessments are detailed installation surveys conducted by local pollution prevention staffs and external pollution prevention experts. The surveys provide each installation with a roadmap for achieving the goals of the program.

Financial Incentives

Changes in the way installations pay for environmental compliance have also provided an incentive for changing internal behavior. Installation commanders are responsible for operating their installations within an operations and maintenance (O&M) budget. Until 1991, waste disposal costs were paid from a central account, not from the commanders' O&M account. Now, all hazardous material, hazardous waste, and municipal solid waste disposal costs are paid from the commanders' O&M account. This change in accounting procedures internalizes these costs to the installation and provides a financial incentive to reduce waste generation. This incentive becomes even more significant when considering that national regulations governing hazardous waste disposal have become more stringent and have driven the unit disposal costs up tenfold over the past five years. Pollution prevention offers commanders an opportunity to reduce the impact of this "must-pay" bill on their budget. This has prompted many installations to manage more carefully their use of hazardous materials, and consequently has reduced the amount of waste generated.

Recent passage of the Federal Facility Compliance Act provides additional financial incentive to minimize hazardous material use and waste generation by removing the federal exemption from some Resource Conservation and Recovery Act (RCRA) requirements. This act now exposes federal installations to potential fines and penalties for notices of violation (NOVs) issued by federal, state, and local environmental jurisdictions for mismanagement of hazardous materials and waste. The rules governing storage, handling, and paperwork requirements for hazardous materials and waste are complex and administratively cumbersome. Approximately 50 percent of all Air Force NOVs are for these types of violations. These represent potential financial liability for having hazardous materials on Air Force installations. Pollution prevention offers an avenue to reduce or eliminate these materials. Organizational accounting procedures that force polluters to pay their own environmental costs, fines, and penalties are a powerful technique for reducing waste generation and improving the corporate bottom line.

Performance Incentives

The metrics for the Pollution Prevention Program are based on installations reporting their baseline figures and updating these figures annually. The data are aggregated at major commands and ultimately at Air Force Headquarters. Scorecards, based on the data, are reported to the Air Force Environmental Protection Committee and rate all commanders on their progress toward meeting the objectives. Each installation, each major command, and Air Force headquarters conduct EPC meetings quarterly. The committees are headed by the commander or vice commander and include key staff from all functional areas. At Headquarters Air Force, the committee is cochaired by the deputy vice chief of staff and deputy assistant secretary for environment, safety, and occupational health. The EPC also regularly advises the chief of staff and secretary on the health of the program and current issues. This EPC provides an opportunity for leadership at all levels to assess the performance of the environmental programs of subordinate levels. Because of the emphasis senior leadership places on pollution prevention, these scorecards have become a basis for competition. Every commander wants to have the best program.

Changing External Behavior

Changing the behavior of suppliers is more difficult. The only control over the private sector is through incentives embodied in the contracting process. This influence can, however, have some effect by virtue of the size of the Air Force budget. For example, in fiscal year 1991, the Air Force executed nearly 5 million contracting actions totaling nearly 50 billion dollars. Throughout the Department of Defense, there were nearly 12.3 million contracting actions totaling more than $150 billion. Purchasing power of this magnitude and the keen competition among suppliers for government business impose a responsibility on the federal government to be an informed environmental consumer.

The Instrument of Change—The Government Contract

The government contract is not only an instrument for acquiring goods and services needed to execute government operations—it is also an instrument for implementing government policy by providing incentives. This is similar to the way the government uses Internal Revenue Service tax codes to encourage specific types of spending and investing behavior.

The Air Force, like the rest of the federal government, buys a wide range of products. This purchase power represents an opportunity to influence the direction of corporate product development efforts toward environmentally preferable performance. These purchases run the gamut from sophisticated, smart weapons systems to commodity items such as office supplies. The procedures for buying these items are similar in some ways but very different in others. One common feature is that each purchase is a contracting action. The only person authorized to commit the government to a contract action is a contracting officer (the actual buyer is someone else). The contracting officer provides a service to the buyer and is responsible for satisfying the needs of that customer. Customers describe their requirements through a technical description of the item or product. If the technical description does not include environmental considerations, they will not be a factor in the purchase. The buyers are a diverse group and the challenge is to educate the buyers to incorporate environmental considerations into their product requirements. To complicate matters, commercial items such as office supplies are purchased centrally by organizations like the General Services Administration (GSA). They buy in quantity and offer products to all agencies of government through catalogs, allowing buyers to purchase items by number through a central contract. Although GSA has introduced a recycled products catalog, it cannot force people to order environmentally preferable products. Under the current system, each office across the government must make sure its own supply organization is ordering the products containing recycled content. This is a classic situation of centralized versus decentralized decision making: Is it better to let each office make its purchasing decisions or to impose them at the contracting office?

On a practical level, there are three sections of a contract for which specific language is needed to put policy into action. The first is Section A of the contract, the "Solicitation Form." Here the Air Force is developing a general statement to be placed on the front page of all requests for proposals (RFPs) describing Air Force environmental policy and philosophy.

The second is Section L, "Instructions, Conditions and Notices to Offerors." For this section, specific formats are being developed that offerors will be required to use to describe the environmental performance of their product, system, or service.

The third is Section M, "Evaluation Factors for Award." It includes two parts called "items" and "factors," which is where the evaluation criteria used to judge the environmental performance of the offeror's proposal must be added. To develop items and factors that use environmental performance as an award selection criterion, several questions must be answered:

1. What attributes or criteria determine whether a product, service, or system has been designed for environment, incorporates the principles of industrial ecology, or demonstrates acceptable environmental performance?
2. How are those attributes or criteria to be measured?
3. Against what standards should the measurements be compared? (This is important because standards establish the minimum performance or compliance level an offeror must meet to satisfy the RFP.)
4. How are the attributes or criteria shown to be cost-effective when compared to "business as usual"?

Once the contractual language is developed that describes what is to be measured, how it is to be measured, and what standards will be applied, it must be incorporated at four stages throughout the contracting action: (1) the request for proposals, (2) the technical review, (3) the best and final offer, and (4) source selection.

Standards can be either quantitative or qualitative. Most people prefer quantitative standards because they are easy and straightforward. An example of a quantitative standard is, "At a weight not exceeding the basic design gross weight, the aircraft is capable of transporting a payload of (a) 30,000 lbs for 2,800 nautical miles (nm) distance, or (b) 48,000 lbs for 1,400 nm distance." The ability to influence the environmental performance of acquisition programs is limited by the ability to prescribe material or process selections or, alternatively, a methodology for making selections. Methodologies for making these selections are not very mature, but they are evolving.

For an organization like the Air Force unilaterally to develop and apply quantitative standards that can withstand legal scrutiny, they must be very specific and narrow in scope, and within the organization's expertise. An example of this type of standard would be "copier paper shall contain at least 50 percent postconsumer waste."

It would be difficult for the Air Force to develop and apply more comprehensive quantitative standards to evaluate the environmental acceptability of a commercially available product, system, or service. To be credible and not appear arbitrary, these types of standards would have to be developed by the EPA, Federal Trade Commission, or some other federal agency or competent government or industry group and would have to be applied to specific products. A national "green stamp of approval," similar to Underwriters Laboratory certification of safety, may work as an evaluation standard.

Qualitative standards may be used in lieu of quantitative standards, but they are more subjective and difficult to evaluate. The following is an example of a qualitative standard extracted from government contracting documents:

> The proposed system environmental program (SEP) will be evaluated for adequacy in effecting the design of changes or modifications to the baseline system to achieve special environmental objectives. The evaluation will consider specific tasks, procedures, criteria, and techniques the contractor proposes to use in the SEP. The standard is met when the SEP
> a. Defines the scope of the system environmental effort and supports the stated environmental objectives.
> b. Defines the qualitative analysis techniques proposed for identifying environmental impacts to the depth required.
> c. Describes procedures by which engineering drawings, specifications, test plans, procedures, test data, and results will be reviewed at appropriate intervals to ensure environmental requirements are specified and followed.

This standard implies a definition of the environmental objectives of the program, an analysis technique, and procedures and means to evaluate their adequacy.

Despite these difficulties a number of efforts are under way to put environmental performance into the source selection process. By narrowing the initial scope of the effort, then using lessons learned and increasing understanding of the industrial ecology concept to broaden the scope over time, the acquisition process can evolve to increasing levels of sophistication.

Today, the start is modest. A standard Data Item Description (DID), and proposal formats to use for describing the environmental aspects of item, system, and service performance, are being developed. In the short term, contractors are being required to list components and end items that incorporate recycled materials. In the long term, changes to the Federal Acquisition Regulations that give preference to recycled products and that integrate environmental performance into source selection criteria may be required.

Buying Commercial Products

The markets for many recycled materials, like paper and glass, have been sluggish recently because the market for the products made from recycled materials is diminishing. This is frustrating Air Force efforts, and those of corporations and municipalities across the country, to increase the percentage of the municipal waste stream making its way back into the raw material stream. This is occurring at the very time municipal landfills are disappearing. In 1978 there were approximately 20,000 landfills in the United States. Today, there are about 4,000. Siting new landfills is becoming difficult and often is an emotionally charged issue in communities. Incineration or waste-to-energy facilities are also highly contentious. The toxic air emissions, the low Btu content of the waste feed, and the potentially hazardous waste resulting from the incineration process make this a much less attractive alternative to recycling the waste into raw material. In addition, siting waste incinerators is difficult. As it turns out, much industrial waste could be reprocessed into raw materials if regulatory, technical, legal, and cultural barriers could be overcome.

While the size of the federal budget in general, and the defense budget in particular, is decreasing, it is still of sufficient size that federal procurement actions can make a difference between survival and failure of fledgling companies that use recycled materials to manufacture useful products. Requiring that purchased products contain recycled material would send a clear message that the private sector's product development efforts should include use of recycled materials, and would stimulate markets.

This is the principle behind one of the Air Force Pollution Prevention Program initiatives, Proactive Procurement, designed to close this material loop. On September 25, 1992, the secretary of the Air Force and the chief of staff signed a policy memorandum establishing an Air Force policy to purchase and use products containing recycled materials when available. Specifically, by the end of 1993, 10 percent of all nonpaper products and 50 percent of all paper products procured by the Air Force shall contain recycled materials. Based on Air Force paper use in the Pentagon alone, this policy will save more than 1,300 cubic yards of landfill space, almost 7,300 mature trees, 3 million gallons of water, 162,500 gallons of oil, and 1.75 million kilowatts of energy, and will eliminate 25,600 pounds of air pollutants each year. The policy also includes lubricating oils, retreaded tires, building insulation, cement, and concrete under the nonpaper category. This policy puts leadership emphasis behind the requirements of RCRA Section 6002, and the EPA affirmative procurement guidelines.

ACQUIRING "GREEN" WEAPONS SYSTEMS: MILSPECS AND MILSTDS

A significant component of the environmental impact of weapons systems results from the use of technical requirements documents, particularly military specifications (MILSPECS) and military standards (MILSTDS). These documents were originally created to standardize the manufacture and maintenance of weapons systems to allow easier logistical support and ensure quality and reliability. Indeed, MILSPECS and MILSTDS became de facto industry standards around the world. The 1989 Montreal Protocol Solvents Technical Options Committee Report stated that approximately 50 percent of CFC-113 used worldwide for the manufacture of electronics circuit boards was driven by U.S. MILSPECS, because they had been adopted as industry standards, and by other governments and organizations such as NATO. Unfortunately, the MILSPECS and MILSTDS system cannot react to market forces as quickly as the private sector. Thus, in rapidly changing areas such as elimination of ODCs, MILSPECS and MILSTDS rapidly become obsolete, hindering rather than encouraging environmental performance of products and manufacturing processes. Significantly, this also impedes the ability of defense industries to change quickly to new manufacturing technologies, an important issue as companies move from defense to commercial business ventures.

MILSPECS and MILSTDS were created at a time when environmental impacts were not thought of or understood. As a result, the documents focus on obsolete notions of reliability and performance rather than environmental concerns. Moreover, the process for changing or eliminating them is far more cumbersome than for creating them. As a result, Air Force and DOD procurement is constrained by bureaucracy that tends to produce an ever-increasing body of requirements documents. The Defense Standardization Program, which formulates, issues, and implements policies and procedures governing MILSPECS and MILSTDS, is managed by the Defense Quality and Standardization Office. They are currently working with the Office of Management and Budget to adopt nongovernment standards in lieu of MILSPECS and MILSTDS. As of 1992, the DOD had adopted about 5,500 nongovernment standards. This trend will help the military services keep pace with the environmental ethic that industry is incorporating into manufacturing activities in response to customer demands, and free defense industries to move their processes closer to those in use in the commercial sector.

Reviewing the experience of the military in phasing out ODCs is a useful study in the dynamics of the MILSPECS/MILSTDS system. Over the past two years, the Defense Standardization Program has automated more than 600,000 pages of MILSPECS and MILSTDS to search for ODC requirements and has looked for opportunities to leverage its efforts to make changes. For example, the program found the one document requiring the use of ODCs in testing procedures for passive electronics devices was referenced by 1,700 other documents. The ODC requirement was eliminated, thus changing the other 1,700. The program also found more than 750 documents that referenced one requirement governing testing procedures for microelectronics. Again, by eliminating ODCs from the one, the program effectively changed the 750.

CHANGING MILSPECS AND MILSTDS

To create a MILSPEC, the preparing organization drafts and coordinates the document in collaboration with affected industries. Once the document is in circulation, anyone can use it. To change a MILSPEC, all the parties who have used it must coordinate on the change. In addition, there are "lead agencies" responsible for specific documents used by others. For example, the Navy is responsible for documents governing the manufacture of electronic circuits and is the only agency that can change them.

Once a MILSPEC is established, however, specific program offices are not bound to apply the specification as written. Because program offices are responsible for delivering the prescribed number of products or weapons systems that meet performance requirements, on time, and within budget, the program offices have the authority to use whatever technical requirements documents they deem appropriate. They can use industry standards, tailor MILSPECS for their program, create their own requirements, or ask the contractor to propose standards and prescribe an evaluation and approval process. The easiest path, and the one that presents no risk to the program office, is to use MILSPECs as written, which is therefore what almost always happens in practice. This is one of the reasons that contractors for major acquisition programs are still being required to adhere to MILSPECs and MILSTDS that require ODCs, despite the Montreal Protocol.

Acting on complaints from defense contractors, Congress added Section 326 to the National Defense Authorization Act for Fiscal Year 1993. The act requires that after June 1, 1993, for any contract over $10 million, any requirement to use ODCs must be approved by the service senior acquisition executive. It goes on to stipulate that this approval can be given only if, after an evaluation, the senior acquisition executive certifies that there is no alternative. In the Air Force acquisition programs, this is the assistant secretary for acquisition.

For new programs, the solution is more straightforward than for existing programs: the contract prohibits ODCs and the program institutes a process for evaluating and making deliberate material choices based on environmental analysis. The Air Force F-22 weapons system program has implemented such a process and is an excellent case study for future programs. The F-22 office incorporated a Hazardous Material Management Program to identify all hazardous materials used for production and maintenance and the specific application for each material. The program office, in cooperation with the contractor, regularly reviews materials being proposed for use and looks for less hazardous alternatives. High on the list of materials to be eliminated are ODCs, heavy metals (such as cadmium and chromium), the 17 chemicals on the 33/50 toxics list, and other chemicals that pose unacceptable risks, such as hydrazine.

For existing programs, the task is more difficult. Contracts have been negotiated and awarded, and MILSPECS and MILSTDS are incorporated into that contractual agreement—the version of the MILSPECS and MILSTDS in effect at the time the contract was awarded. Changes to MILSPECS and MILSTDS that occur after the contract was awarded do not apply. This situation poses problems both for the program office seeking to change the selection of materials in manufacturing processes and for the contractor seeking to implement changes. A major acquisition program typically involves hundreds of individual contracts. Contractors typically use a single manufacturing process to produce parts for more than one program and often for programs belonging to more than one service. This means a contractor wanting to change a process must identify each program and contract affected, individually convince each program office that the proposed change is technically sound, and negotiate a change to each contract with each contracting office. Each of these steps is labor intensive.

Probably one of the most difficult steps is convincing the program office that the proposed alternative is technically sound. Each program office has the authority to accept or reject the proposal, and demand any level of technical data from the contractor it deems necessary. After all, it is the contractor who is asking for approval. Moreover, there is a cultural dimension that influences the adoption of environmentally preferable alternatives. MILSPECS and MILSTDS have defined state of the art for quality and reliability for a long time throughout industry. Although it is dangerous to generalize, there is a tendency for organizations preparing the standards and for program offices not to consider private-sector experience in the commercial market with environmentally preferable technologies to be an appropriate basis for considering military use of the same technology. In support of this view it has been argued that military equipment is used in more harsh environments, or that the military needs better reliability. There may be some merit to these arguments for some applications, but the commercial experience cannot be discounted totally. When large, established corporations make significant investments in environmentally preferable manufacturing technologies, they do so with a confidence the products manufactured with the technology will perform reliably in the marketplace. Nevertheless, this attitude is a factor that must be considered.

One way to accelerate the approval process is to shift the burden of proof. Rather than require the contractor to prove that the proposed alternative performs better than the current process, let the program office show cause why the alternative is inadequate. For ODCs this is exactly the effect that Section 326 of the Defense Authorization Act for Fiscal Year 1993 will have on the approval process. For any contract requiring ODCs, the senior acquisition executive must certify that no alternative exists, based on technical evaluation. This forces program offices to certify up through command channels to the assistant secretary why alternative technologies will not meet performance requirements. Most military use of ODCs is for solvents, and based on the conclusion of the most recent United Nations Solvent Technical Options Committee, there are alternatives for virtually all of the solvent uses it identified. Today many major weapons systems development and production programs would come to a halt if the supply of ODCs to contractors were suddenly stopped. One small step that would help those contractors eager to move toward environmentally preferable alternatives would be to allow them to adopt MILSPEC and MILSTD changes enacted since the contract was awarded simply by notifying their program offices, without the need for approval.

To move toward environmentally preferable alternatives, program offices must be receptive to new design ideas, manufacturing processes, material choices, and products. The process for approving environmentally beneficial changes to programs must be streamlined.

THE NEED FOR TECHNOLOGY ASSISTANCE

Program offices tend to have relatively small staffs, and as a result, most of the technical expertise resides with the contractor. This places the program offices at a distinct disadvantage when trying to evaluate the attributes and risk of a contractor-proposed, environmentally preferable alternative technology. By definition, contractors and their government customers are adversaries. One way of relieving the program office of some of the burden of risk of change is to establish a means of validating environmentally preferable technologies for military use. This could be done by abandoning MILSPECS and MILSTDS in favor of industry standards, or by accelerating the process of changing them and giving environmental considerations a high priority. There are many research and development resources available for this mission, including national laboratories. The key is institutionalizing the means of communicating the endorsed technologies to the program offices and the organizations that prepare MILSPECS and MILSTDS across the services.

SUMMARY

The federal government has an opportunity and a responsibility to help improve the global competitiveness of U.S. defense industries (and other U.S. industrial sectors as well). As defense industries look toward commercial markets to replace defense business in a climate of decreasing defense budgets, the government can stimulate improvements in the environmental performance of manufacturing process technology and of manufactured products in a variety of ways. As a large consumer, the federal government should be an informed and responsible environmental consumer and use contracting incentives to stimulate markets for environmentally preferable products. As a large and sophisticated research organization, the government can provide new seed technologies for the private sector to incorporate into environmental products and services for a global commercial market. As a developer, builder, and owner of high-technology systems, the government can support the integration of environmentally preferable technologies into the U.S. manufacturing base and stimulate a better understanding of the environmental impact of material choices for processes and products. As a major consumer of commercial products, the government can stimulate markets for recycled products and stem the flow of waste to landfills. And as operator of major installations that are, in effect, small cities, the government can reduce the environmental impact of its internal operations and provide a model for municipalities nationwide in their effort to improve environmental quality.

The Air Force Pollution Prevention Program is comprehensive and multidisciplinary and has a heavy technology focus. It exploits opportunities to choose environmentally preferable alternatives presented by each of the roles it fulfills in society. It also exploits the opportunities government has to find and support environmentally preferable choices through a comprehensive program of educating team members about the opportunities presented by their jobs, establishing internal goals and objectives designed to improve the environmental performance of Air Force operations, instituting a series of metrics and scorecards to measure performance, and implementing an investment strategy. Goals are defined and measurable and, to the extent possible, reflect a systems-based approach. It will thus be possible to measure progress over time. In this way, the Air Force offers a model for applying industrial ecology concepts in the short term across a complex organizational structure.

The Greening of Industrial Ecosystems. 1994.
Pp. 165–170. Washington, DC:
National Academy Press.

Designing the Modern Automobile for Recycling

RICHARD L. KLIMISCH

Public opinion polls consistently show a high percentage of the U.S. population, typically more than 75 percent, consider themselves to be environmentalists. One of the issues that environmentalists feel most strongly about is recycling. In many communities there is an almost religious fervor surrounding recycling. Unfortunately, there also appears to be widespread misunderstanding regarding the extent to which recycling can contribute to solving our environmental problems and how far we have already gone with recycling. For example, it is not generally known that motor vehicles are already among the most widely recycled products in the marketplace; 75 percent, by weight, of vehicle materials are currently being recycled. The recycling rate for aluminum beverage containers is around 60 percent. There is a very well-developed infrastructure in existence that efficiently recycles essentially all the metallic components of vehicles.

This case study will describe the current state of recycling of automotive materials in the economy and some of the trade-offs among environmental benefits that must be faced to achieve increased recycling.

Recent successes have been achieved with lead-acid vehicle batteries (achieving 95 percent lead recycle rates, thanks to the efforts of Battery Council International), with chlorofluorocarbons used in vehicle air conditioners, and with coolants (recovered by automobile dealerships). There is also progress in dealing with tires, oils, and other materials consumed during operation of a vehicle. The major focus, however, of a new organization dedicated to increasing the recycling of automobiles, the Vehicle Recycling Partnership, formed recently by General Motors (GM), Ford, and Chrysler, will be the postconsumer fate of the nonmetallic components of the vehicle body and powertrain. At present, there is a high recy-

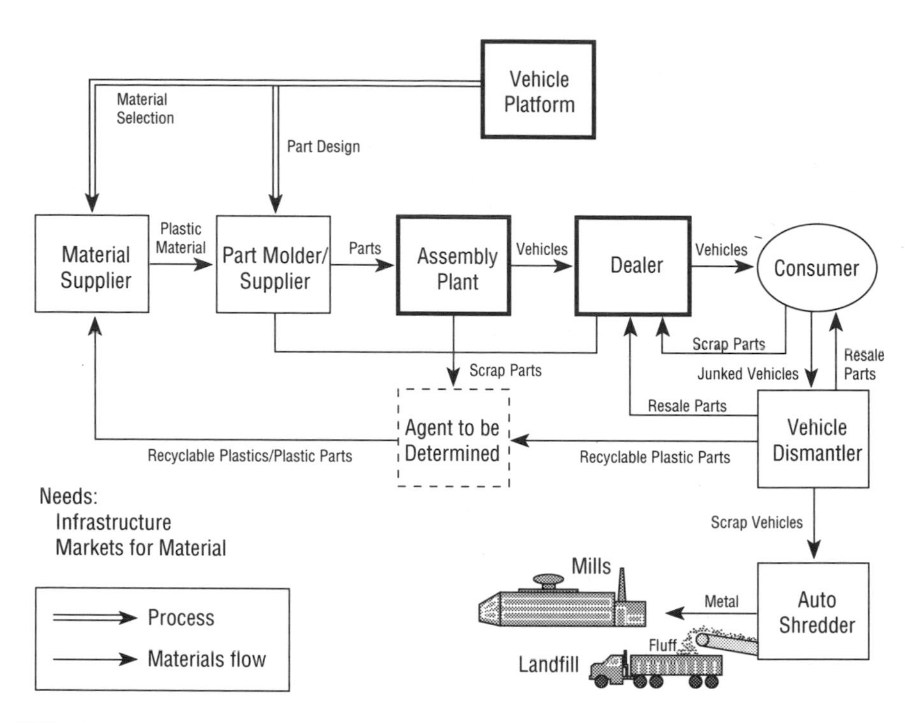

FIGURE 1 Recyclable vehicle process.

cling rate for metallic vehicle components but a low recycling rate for the nonmetallic materials.

In considering this issue, it is useful to begin by looking at current practices (see Figure 1). The first link in the recycle chain is the dismantler, who sells parts from vehicles either directly to customers or to remanufacturers. Some of the original equipment manufacturers (OEMs) are directly involved in remanufacturing. For example, GM recovers engines from dismantlers to produce remanufactured parts for out-of-production engines, and other (non-OEM) companies are involved in the remanufacturing of a variety of automotive parts. Most dismantlers allow the hulk to remain on their property to allow "chrome pickers" to remove and purchase miscellaneous parts. After some period of time, based on demand and space, the dismantler sells the hulk to a shredder. There the metallic materials (iron, aluminum, copper, etc.) are separated by magnetic and other methods, leaving behind a mixture of plastics (30 percent), glass (20 percent), rubber (15 percent), and dirt (25 percent). Each car that is shredded yields an average of 600 pounds of this residue, called fluff. Virtually all the fluff is currently being landfilled (in the United States, and elsewhere). Fluff is only a small part of total municipal solid waste. In the United States, 4.4 pounds of waste are buried each day for every person in the country: one ounce of this daily waste is automotive fluff. This amounts to a total of 3 million tons of fluff per year in the United States.

The Vehicle Recycling Partnership (VRP) has divided its efforts into three groups: the Shredder Residue Group, the Disassembly Group, and the Design Guidelines Group. The word *recycling* is somewhat unfortunate and misleading in the title of the partnership because its real goal is "to minimize landfilling of vehicle materials." In some cases, maximizing recycling will clearly be a suboptimization environmentally and economically. Furthermore, expectations for 100 percent recycling are as unrealistic as are those for closed-loop recycling and for zero waste; it is a goal, but unachievable in many cases.

The Design Guidelines Group has as its mission "to develop material selection and design guidelines to facilitate reuse, recycling, or reclamation of materials and components from post-consumer vehicles." These guidelines will lead to designs that facilitate recycling, as opposed to current vehicle designs that do not lend themselves to recycling. Initiatives that will be or have been taken include marking plastics by type to facilitate separation and recycling, reducing the number of resins used, developing environmentally appropriate attachment and detachment techniques, and improving material compatibility. The most intractable situations today involve the joining of incompatible parts by adhesives, which is often done, for example, for bumpers and instrument panels. It is important to combine these efforts with the concomitant efforts to improve design for manufacturability. Obviously, it is also important to involve the suppliers, the dismantlers, and the shredders in this effort, and negotiations are under way to develop cooperative arrangements with these various groups. In addition, there are agreements with government laboratories supporting related studies.

The two other VRP work groups seek to improve the recycling rate for both current and carryover designs, which involve vehicles that are already on the road and will dominate the automotive disposal situation for the next 10 to 20 years. The Shredder Residue Group will focus on efforts to use the existing infrastructure that optimally recycles iron (the material for which recycling markets currently exist) "to reduce the total environmental impact of automotive shredder residue through socially responsible, economically achievable solutions by resource recovery or secondary uses." The group will, for example, examine techniques to separate the various materials in the fluff, perhaps through selective solvent dissolution (as pioneered by Argonne National Laboratory). Making plastic materials recyclable both in themselves and in different configurations will also be an area of great interest for this group.

The mission of the Disassembly Group is "to enable the efficient disassembly of components from vehicles that have value for reuse, to recover valuable materials for recycling, and to remove potentially harmful materials prior to becoming auto shredder residue." It will focus on creating an infrastructure based on separation of materials at the disassembly stage. The marking of plastic parts through a standard called SAE J-1344, which has been adopted by the major vehicle manufacturers, is designed to facilitate this sorting of materials (in fact the sorting would be extremely difficult without it). These efforts will also involve the dismantlers as well as the materials and parts suppliers.

One approach under consideration at most of the automobile manufacturing companies in Germany and the United States would require the suppliers, as part of the initial contract, to retrieve and recycle the materials in their products at the postconsumer stage. The role of the OEM would be to use some of the recycled materials, but the OEMs are going beyond this and will help set up the infrastructure and markets. This is important because of their specialized expertise. The infrastructure for the recycling of plastics or rubber, for example, is likely to be different from that for metals. As pointed out by Frosch and Gallopoulos (1989), the metal recycling industry is limited by demand considerations (supply is adequate because the magnetic and other properties allow easy separation) while plastic recycling is typically limited by supply considerations (as separation of specific polymers from the fluff, for example, is much more difficult). But because increased vehicle recycling is based on an infrastructure that has been optimized for iron recycling, it will undoubtedly require significant additions to the infrastructure. If separation is accomplished at the dismantling stage, there are likely to be intermediaries to collect, clean, and prepare the materials for the suppliers. The separation of nonmetallic materials will be a major focus of all three groups.

German industry has estimated that separation, collection, cleaning, and grinding of plastics will cost more than $0.30 per pound, enough to make recycling economics problematic for inexpensive commodity plastics such as polypropylene or polyethylene. On the other hand, the economics for many of the more sophisticated engineering plastics, which cost more than $2.00 per pound, appear favorable. Obviously, these numbers are extremely sensitive to disposal costs. Ultimately, if it costs less to make new parts from virgin material than from recycled materials (even after internalizing disposal costs), other options, including new technologies, must be considered. Thus, for example, the environmentally preferable option for many plastics might be as an energy source, rather than a recycled material. There are active projects in GM and elsewhere to study various thermal recovery techniques. For example, GM has an active program on pyrolysis, and Chaparral Steel and other shredders have been attempting to obtain permits to use fluff as a fuel in cement manufacture. Others favor straight incineration for energy recovery. The rationale for such projects is that these materials are derived from petroleum so that thermal recovery offsets the use of virgin petroleum to make new parts, which can be made from "virgin" petroleum more economically. The cement kiln is especially attractive as a variety of the fluff components, including glass, rubber, and residual iron can be used as energy and material inputs for cement manufacture. Social acceptance of such options, however, will be difficult.

INDUSTRIAL ECOLOGY AND LIFE CYCLE ANALYSIS

If one looks closely at the concept of cradle-to-grave responsibility for products through life cycle practices, it is apparent that design changes to enhance recyclability can cause problems elsewhere in the life cycle of the product. One of the simplest examples involves substituting steel for plastic parts on an automobile. This certainly enhances recyclability but, because it increases weight, it causes more fuel to be consumed during the use phase of the vehicle. Given the long use phase of modern automobiles, the net environmental impact of such a substitution will most likely be negative. Similarly, advanced composite materials are generally more complex than the materials they replace. This makes recycling more difficult, but the lower weight and the better durability of such advanced materials yield environmental benefits that more than offset the negative environmental consequences of postconsumer disposal of such materials. The trade-offs at the disposal, manufacturing, and use phases are extremely complex, involving, for example, trying to balance different kinds of pollution, different end points, different subjects, and different locations. Somehow the designer needs to comprehend the whole life cycle and take account of these extremely complex trade-offs. When one considers how difficult it has been to decide whether polystyrene or paper cups are more environmentally correct, this would seem to be a tall order for products as complex as automobiles. The designer must also consider trade-offs among recyclability, cost, and performance. Obviously, it is necessary to find some way to simplify the analysis for the designer. In this regard, VRP is developing simple design preference guidelines, which provide a first step in working through this problem.

REGULATORY ASPECTS

One of industry's greatest concerns is that the government will intervene and dictate answers to these difficult problems by means of the conventional command-and-control regulatory approach. The concern is that the application of inflexible regulations on product design will harm the competitiveness of the domestic industry without improving—or possibly even degrading—the environment, as mentioned above. The problem with such regulations is that they do not allow the flexibility to experiment with different approaches and technologies. They also tend to pit the producers against their customers. By providing little consumer incentive, such regulations tend to be difficult to implement.

Furthermore, as with corporate average fuel economy, or CAFE, standards, such regulations are fraught with differential competitive implications. If, for example, Congress dictates inflexible command-and-control recycling regulations, research and development on advanced materials would be seriously curtailed. Further, inflexible recycling regulations that fail to consider the life cycle trade-offs may well have a net negative effect on the environment. The problem is compounded by the lack of technological expertise among environmental regulators.

On the other hand, the good news is that this issue—green design in general—has great potential to cause a fundamental shift in environmental regulation as well as in material management. The clear need is for flexible regulations that create harmony between the manufacturer, the government, and the customer. Only market-based regulations can create such harmony.

SUMMARY

In the final analysis, industrial ecology considerations, life cycle waste management, and pollution prevention are quite similar. They are based on practices that, in the automotive industry, are already in place. Ultimately, these approaches must be combined with total quality management to meet the future demands for "greenness," which we all expect our future customers to require. One of the most difficult barriers to the success of such comprehensive approaches is the inflexible, non-market-based, command-and-control, environmental regulatory system now in place.

REFERENCE

Frosch, R. A., and N. E. Gallopoulos. 1989. Strategies for manufacturing. Scientific American 261(3):144–152.

The Greening of Industrial Ecosystems. 1994.
Pp. 171–177. Washington, DC:
National Academy Press.

Greening the Telephone: A Case Study

JANINE C. SEKUTOWSKI

AT&T is currently conducting a demonstration project called Green Product Realization to provide a basis for a more comprehensive "green" design program. The demonstration project will be used as a learning experience to generate feedback about the relevancy and utility of potential design for environment (DFE) guidelines and tools and to explore all elements of a system for delivering "green" products to customers.

The DFE effort at AT&T represents one of the first attempts to link industrial ecology concepts to specific industrial practices. The challenge of implementing DFE lies in the extreme difficulty of quantifying environmental attributes in a way that permits comparison among environmental effects and life cycle stages. Moreover, there are many data to which AT&T lacks access, such as information about the environmental effect resulting from the mining or processing of materials purchased by the firm.

At AT&T, "Green Product Realization" is the designation of a program for developing the full-stream capability for minimizing the adverse life-cycle environmental impacts of manufactured products throughout their entire life cycle (see Figure 1). It builds on a concurrent engineering program used by AT&T and many other companies. This system is called DFX, where DF stands for "design for" and X stands for downstream (from design) considerations such as manufacturability, installability, reliability, or testability. DFE, therefore, is any product or process design activity that minimizes the adverse environmental effects of the manufacture, use, and eventual disposal of products (see Figure 2).

However, although DFE principles may be applied to produce a "green" product, if, at the end of its life, the product ends up in a landfill, can it still be consid-

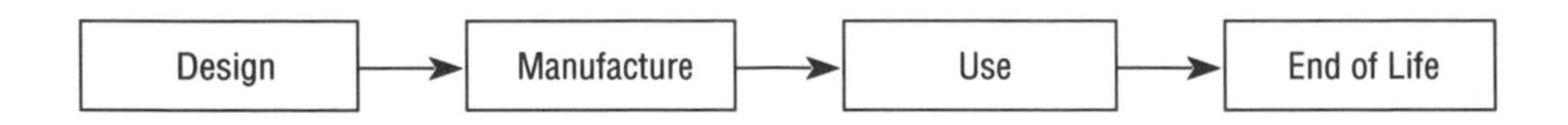

FIGURE 1 Simplified product life cycle.

ered "green"? It is important to extend our concern for minimizing the environmental impact of products beyond the traditional boundaries of design and manufacture to include such downstream considerations as use and end-of-life. It is in part to understand the implications of this paradigm shift that AT&T began a study to investigate what makes a product "green" from a life cycle perspective.

To speed the development and application of "green" design principles, and to investigate the environmental implications of the entire life cycle of a product, a single product was selected for the demonstration project. The criteria used to select the product were its size, number of components used, and length of production cycle. The obvious small product, with a minimum number of components and with a short product delivery cycle, given AT&T's product mix, was a telephone. This led to the Green Phone Feasibility Study Project.

DEFINING "GREEN"

For this feasibility study, a narrow view of environmentally preferable options, or "greenness," was selected. One reason for this is that the applicable methodologies, such as life cycle analysis (or LCA), are not very rigorous at this time. AT&T felt that it could not wait to begin developing "green" product principles until methodological questions were settled. For purposes of this project, for example, if a recyclable plastic is used, this is viewed as "green," even though, if a rigorous life cycle analysis were done, it might not be truly "green" when balanced against the energy consumed and the waste produced to manufacture and recycle the plastic.

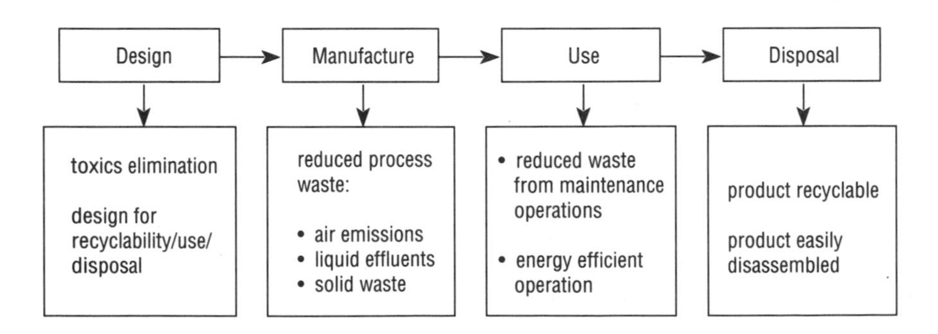

FIGURE 2 DFE product/process design activities.

Another attribute defined as "green" is the use of options that make for easy recycling of materials, such as

- Applying the rules of design for assembly, because it implies design for disassembly (i.e., a product that is designed for assembly is generally easier to disassemble into component pieces for recycling).
- Labeling plastic parts to facilitate separation after disassembly.
- Using no metalized plastics.
- Avoiding use of different plastics in the same separable component.

Other practices defined as environmentally preferable included the following: (1) waste from manufacturing is not simply landfilled, and (2) the product is not landfilled at the end of its life.

ORGANIZING THE GREEN TEAM

AT&T assembled a cross-functional team for this feasibility study. The team includes representation from product management and marketing, design and manufacture, R&D, and the corporate environmental organization. One of the first activities was a "green" product special interest group breakout session at a users group conference. In general, the "green" product concept was well received by customers. When asked if they would purchase a "green" product rather than another product, price and function being the same, all but one customer responded yes. All of the customers said that they would consider purchasing a product with refurbished or remanufactured components, provided the quality was as good as a "new" product, although the majority expected the price to be lower. Ninety percent of those polled said they would participate in a recycling program, and all of those polled said they would accept biodegradable packaging and documentation printed on recycled paper.

With such a favorable response from customers, the team decided to determine how "environmentally preferable" AT&T's current processes are, that is, to establish a baseline for telephones currently being manufactured. The baselining task was divided into the four product phases: design, manufacture, use, and end-of-life.

GREEN BASELINE

Design

To help evaluate the environmental aspects of the design of the telephone, the AT&T team is developing a "green" scoring system, which assigns numerical points for various environmentally preferable attributes. The design is evaluated on how well it meets these attributes. For example, if the design possesses a particular "green" attribute, it would get a score of five, whereas it might get a score of three for partial attributes. The final "green" rating is then calculated from the ratio of the actual scores and the maximum possible score; a perfectly "green" product—based on the selected attributes—would receive a rating of 100. This calculation is a convenient method for comparing design options.

A software-based tool to calculate the "green" score of a product design is being developed. Such a tool would provide an easy, uniform way to assign numerical points to various environmentally desirable attributes. It would be a useful way of evaluating the environmental effect of potential design changes. It would also allow evaluation of the "greenness" of a product family and could be used to track progress in each successive generation of product.

The scoring system for "greenness" of design includes the following product attributes:

- Can the material variety be minimized?
- Can components be reused or recycled?
- Are recyclable plastics used?
- Can metal parts be detached and recycled?
- Are plastics identified with established standard ASTM (American Society for Testing and Materials) or ISO (International Organization for Standardization) marks for recycling?
- Can the product be easily disassembled?
- Are screws used?
- Are there any hazardous elements?
- Are heavy metal pigments used?
- Are brominated flame retardants used?
- Are adhesives used?

Based on these attributes, the score for the design of the telephone was quite good; in fact, the existing design of the telephone already incorporates a number of environmentally desirable elements. Apparently, this is mostly due to the fact that a rigorous DFA (design for assembly) analysis was done on the design of this telephone. There are, however, some improvements that can be made to make the design of the telephone "greener":

- Materials in components and housings can be identified with ASTM or ISO markings for easier recycling.
- It is possible to reduce somewhat the variety of materials used; for example, the same grade of ABS plastic should be used throughout.

Going forward, AT&T hopes to incorporate these findings in the next generation of products to make them even more environmentally preferable. This will be aided by improvement to the scoring system as LCA and DFE methodologies are developed further.

Manufacture

Another stage of the life cycle of the telephone for which a baseline must be established is, obviously, the manufacturing stage. For purposes of the initial analysis, an initial boundary condition of considering only those components manufactured in AT&T factories for the "green" evaluation was established. This somewhat arbitrary restriction was necessary for practical reasons: data from hundreds of suppliers simply were not available and would in many cases be proprietary. Although it is difficult, at this time, to conceive whether or how suppliers (original equipment manufacturers, or OEMs) should be evaluated for "greenness," it may be appropriate to consider them at a later date. In the interim, AT&T has used generic tools, such as contract and purchase order provisions, to impose clearly desirable baseline environmental standards on suppliers (such as not using packaging made with chlorofluorocarbons).

The following qualitative statements can be made about the "greenness" of the manufacturing process for a telephone:

- A significant quantity of waste is generated in the packaging of components that are assembled into the telephone.
- Similar waste streams are treated differently at various factories.
- Some of these waste treatments are environmentally preferable to others (why these preferable methods have not been adopted throughout the firm will be analyzed in the future).

The next task is to analyze quantitatively the manufacturing of the telephone. The entire manufacturing sequence will be characterized with an objective of identifying and quantifying all wastes (air, liquid, and solid) and all the associated costs and waste handling options. A scoring system, similar to that developed for the design phase, may then be developed for manufacturing to calculate "greenness" and track progress. Because of the complexity of the manufacturing process, this task will be extremely difficult. For example, frequently several products are made on the same manufacturing line using the same inputs, making it difficult to quantify wastes for any one product type.

Use

For purposes of this initial investigation, the AT&T team is assuming that the "greenness" of the use phase is good and no further characterization is necessary for two reasons. First, the energy consumption of the telephone is minimal. Second, the environmental impact of the design and manufacturing, and perhaps end-of-life, phases is many times greater. It is also possible that some telephone use substitutes for energy-intensive transportation, making the telephone as it functions within our economy an environmentally preferable product. Thus, the use phase will be examined for "greenness" at a later stage of the project.

End-of-Life

Work on evaluating the end-of-life phase of a telephone for "greenness" has not yet begun. However, there are several observations that can be made.

A recycling infrastructure for AT&T telephones is available. This infrastructure is a legacy from the Bell System and has existed for many years. It used to be that the Bell System leased telephones to customers. These telephones were designed to last for 40 years and it was only after this length of time, or if the customer moved, that the telephones were returned to the Bell System. These telephones were then refurbished and leased to different customers, or AT&T factories "mined" the returned telephones to recycle or reuse various materials. In essence, AT&T had a "take-back" policy long before such policies became popular with environmental regulators.

This AT&T recycling infrastructure is still in use for leased products. However, the percentage of leased telephones is declining for two reasons. First, customers are now buying telephones rather than leasing, and these telephones are only sporadically returned to AT&T. Second, because technology is changing rapidly and new features are continually designed into telephones, customers are buying telephones more frequently to take advantage of these new features.

For leased products, the "green" score for the end-of-life stage would be high; it is unclear what the score would be for products that are sold, because it is not known what happens to them. AT&T will investigate an integrated return policy for its telephones. Unlike the Bell System return policy, this one must be economical to support the reality of being in an unregulated, highly competitive business.

THE NEXT STEPS

As the demonstration project proceeds, several tasks require further effort. For example, as a result of the project several technological options for designing a "greener" telephone have been uncovered. These options need to be considered in the design of the next generation product. DFE guidelines for telephones must be developed and implemented to ensure a uniform approach to environmentally preferable design. The "green" design scoring system needs to be further developed and extended to other life cycle stages. An economical, realistic take-back program for telephones needs to be explored.

AT&T hopes to apply the lessons gained from the demonstration project to other products as well. In addition, as "greener" materials and process technologies become available, they need to be implemented. An example is the development of lead-free interconnection technologies, such as conductive adhesives or lead-free solder alloys. There are, however, substantial data and methodological gaps that AT&T, acting independently or with a few other progressive companies, cannot bridge. Development of international or national priorities and standards for material selection would be useful (if done in an open process free of the temptation to craft trade advantages out of environmental standards).

The development of environmentally preferable products is an evolving process, and it might now be said that work on a "greener" telephone has begun. A truly environmentally preferable telephone is the ultimate goal.

The Greening of Industrial Ecosystems. 1994.
Pp. 178–190. Washington, DC:
National Academy Press.

The Utilization-Focused Service Economy: Resource Efficiency and Product-Life Extension

WALTER R. STAHEL

A key difference between the industrial economy and the service economy is that the first gives value to products that exist materially and are exchanged, whereas value in the service economy is more closely attributed to the performance and real use of products integrated into a system. In our classical, industrial economy, the value of products is essentially identified with the costs of producing them, whereas the notion of value in the service economy is shifted toward the evaluation of costs incurred to provide results in use.

The first approach considers the value of a personal computer with a printer. The second, on the other hand, evaluates the actual performance of the system, taking into consideration not only its cost of production but also all sorts of costs associated with successful use (such as the cost of learning to use it and the cost of repair and maintenance) as well as the quality of the result. In the service economy, what is purchased is the *functioning* of a tool; people buy "system functioning," or performance, not products.

CHANGES AND OPPORTUNITIES AHEAD

Reuse and recycling are among the strategies for waste minimization that should lead us (i.e., the industrial countries that represent 20 percent of the world population but consume 80 percent of all resources) to a more sustainable and resource-saving economy. But how?

Waste-reduction and resource-saving strategies can be applied to the following activities:

- Production (clean technologies through closed loops—a goal of zero waste, primary materials recycling).
- Use (long-life goods, product-life extension, and more intensive use of goods and systems).
- Postuse (secondary recycling of "clean," or sorted, materials, and refurbished components and parts).

Closed loops are one common denominator of waste reduction in these three radically different areas. The other common denominator is the need for management decisions to implement waste prevention. Waste prevention does not happen by accident!

There are two kinds of loops that differ fundamentally with regard to their feasibility: (1) reuse of goods, and (2) recycling of materials (see Figure 1 and Table 1).

The reuse of goods means an extension of the utilization period of goods, through the design of long-life goods; the introduction of service loops to extend an existing product's life, including reuse of the product itself, repair, reconditioning, and technical upgrading; and a combination of these (Figure 2). The result of the reuse of goods is a slowdown of the flow of materials from production to recycling (or disposal). Product-life extension means waste prevention not only through increased use but also in production, distribution (including packaging), and recycling/disposal, as well as a reduction of the environmental impairment caused by the transport necessary for these activities. Reusing goods and product-life extension imply a different relationship with time.

The recycling of materials means simply closing the loop between postuse waste (supply) and production (resource demand). Recycling does *not* influence the speed of the flow of materials or goods through the economy.

It can be shown that there are fundamental differences in the economic feasi-

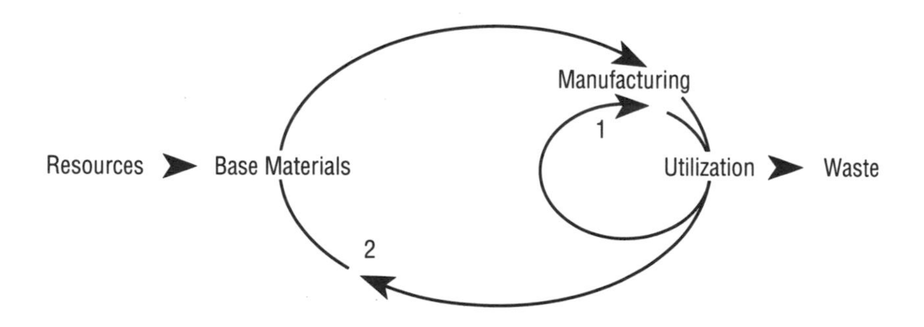

FIGURE 1 The reuse loop and the recycle loop. Loop 1: Waste prevention, long-life products, and product-life extension. Loop 2: Waste reduction and recycling of materials. SOURCE: Stahel and Reday (1976/1980).

TABLE 1 Fundamental Differences Between Reusing Goods and Recycling Materials

Reusing goods through product-life extension activities such as reconditioning	Reusing raw materials through resource recycling
Potential sales price depends on: • Prices of comparable new goods (e.g., bottles, tires, cars) on the local markets • Utilization value • Rarity and collection values	Potential sales price depends on: • Commodity prices on the world market (e.g., scrap steel, sand for glass, rubber) • State of the world economy and of economic cycles
Characteristics of potential buyers are: • Large number (everybody) • Small volumes needed • Short transport distances to buyer	Characteristics of potential buyers are: • Small number (monopolistic markets) • Substantial volumes demanded • Long transport distances to buyer
Intensity of activity • Mainly skilled labor	Intensity of activity • Mainly energy
The following examples will explain these differences in concrete terms, comparing the reutilization of tires with the reutilization of the raw materials inherent in tires:	
• Retreading using a special "cold" process that enables a multiple repetition of the process • The process is environment friendly, uses medium temperatures (90°C) and produces no waste	• Recycling using a special process that consists of freezing the tires with liquid nitrogen and subsequent shredding. Irreversible • The process is environment friendly and produces no waste

SOURCE: Stahel (1986).

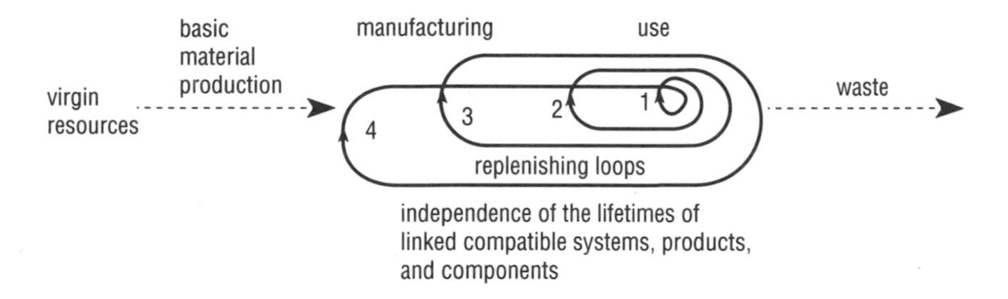

FIGURE 2 The self-replenishing system of product reuse and recycling services. Loop 1: Reuse of goods. Loop 2: Repairs of goods. Loop 3: Reconditioning or rebuilding of goods. Loop 4: Recycling of raw materials. SOURCE: Stahel and Reday (1976/1980).

bility of these two loops. One important difference is that the smaller the loop, the more profitable it is (Table 1). It could therefore be expected that in a free market economy, manufacturers, guided by the "invisible hand," would jump to develop opportunities for reusing goods, without even looking at recycling. We all know that, in reality, the exact opposite is happening. There are gains to selling the use of goods instead of the goods themselves. When we have to sell the use of goods, we will have at least the same turnover and profit as before, when we sold products, once we made the adjustment. Again, what about the invisible hand that seems to have gone astray?

You might object that this is due to the fear that, if we considerably extend the useful life of goods in our consumer society, the economy will break down. The contrary, however, is true. Product-life extension is in many cases a substitution of labor for energy (Figure 3 and Table 2). The missing link in our analyses is *liability*:

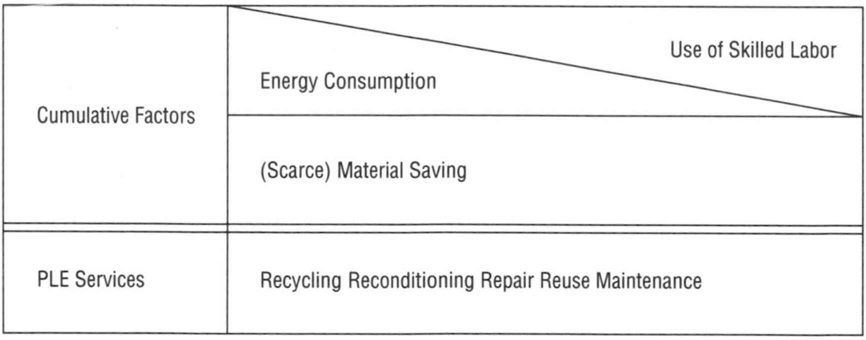

Cumulative Factors	Energy Consumption / Use of Skilled Labor
	(Scarce) Material Saving
PLE Services	Recycling Reconditioning Repair Reuse Maintenance

FIGURE 3 The trade-offs among energy, labor, and materials for each step in product-life extension (PLE). SOURCE: Stahel and Reday (1976/1980).

TABLE 2 Comparison of Labor and Energy Input for Production and Maintenance of a 10- and 20-Year Lifetime Car (excluding operation)

For the Average European Car	10-Year Lifetime Car (Today)			20-Year Lifetime Car (Possible Future)		Change from 10- to 20-Year Lifetime Car	
	1 per car	2 per car/yr	3 Δ*	4 per car	5 per car/yr	6 per car	7 per car/yr
Energy consumption	(in Toe)	(in Toe)		(in Toe)	(in Toe)	(%)	(%)
In basic materials (57%)	0.85	0.85	20	1.020	0.051		
In manufacturing (43%)	0.65	0.065	10	0.715	0.035		
Total Energy Consumption (transport during production phase was ignored as it represented a low %, i.e., 3 to 4%)	1.5	0.150	16	1.735	0.067	+16	–42
Labor	(in man years)	(in man years)		(in man years)	(in man years)	(%)	(%)
In basic materials	0.03	0.003	10	0.033	0.0016		
In manufacturing	0.11	0.011	20	0.132	0.0066		
In production	0.14	0.014		0.165	0.0062	+18	–41
In maintenance and repair	—	0.020	50	—	0.030		
In reconditioning	—	—		—	0.015		
Total labor	—	0.034		—	0.0532		+56

NOTES: Toe = tons of oil equivalent; Δ* = % increase of column 4 over column 1.

SOURCE: Stahel and Reday (1976/1980).

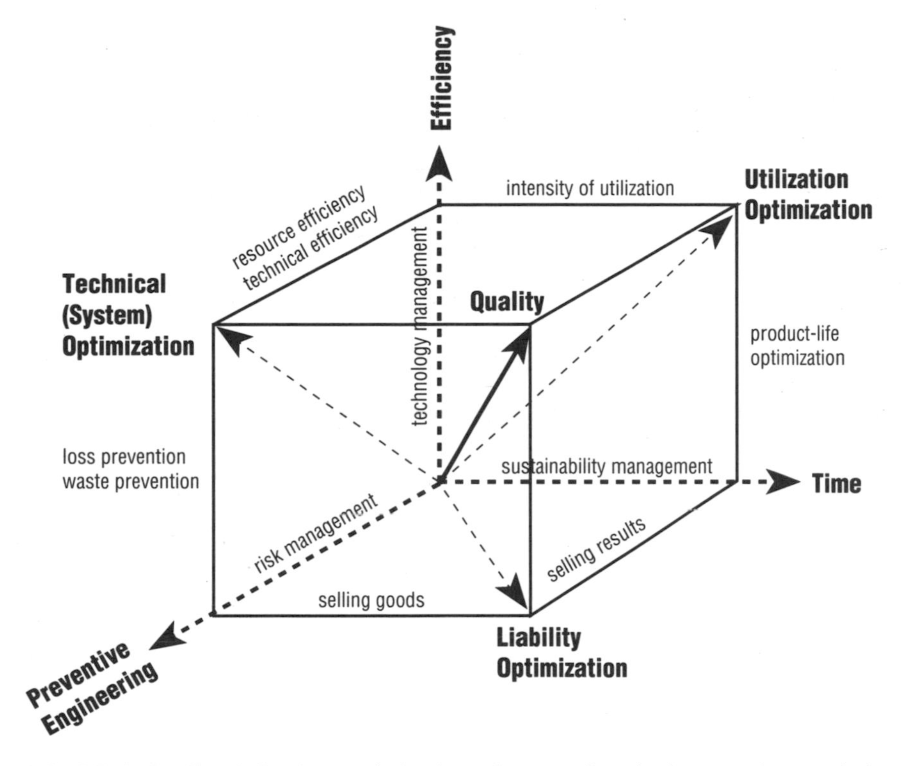

FIGURE 4 Quality defined as optimization of system functioning over long periods of time. SOURCE: Stahel (1991).

- The logic of "reuse loops" implies perfect quality and a broad service liability.
- Selling only the use of goods implies an unlimited product liability by manufacturers and their agents (importers and shops), "from cradle back to cradle" including an environmental impairment liability at all levels (Figure 4 and Tables 3a and 3b); this is also inherent in the German Waste Law of 1986 and its provision (para. 14) regarding the manufacturer's obligations to take goods back and recycle them after use.

RECYCLING

The nice thing about the recycling of materials is *not* its economic attractiveness, but the fact that it carries a limited product and material life-cycle liability. The liability "from cradle to grave" that is so fashionable today is really "for cradle *and* grave"; it excludes unlimited use liability as well as the obligation to reuse

TABLE 3a How Selling the Utilization of Products Rather than the Products Themselves Affects Liability of Economic Actors

	Alternative Utilization and Commercialization Opportunities	Examples of Economic Actors	Alternative Liability Carriers "product quality" and "utilization"		
			Producer	Fleet Manager	User
SALE	Owner is user Durable product is sold (cash or credit) • cars • white goods • clothes • computers	Individuals	Risk of warranty (6/12 months)	—	All risks for an unlimited time period (except warranty)
RENTAL	Owner is fleet manager Product is rented • cars System is rented • apartment • hotel room Service is rented • taxi	Companies and individuals offering service • shops • Hertz, Avis • investors • hotels • taxi owners	Risk of warranty	All risks for the full utilization period (except warranty)	none
SELLING SYSTEM UTILIZATION	Owner is fleet and maintenance manager System utilization is sold • transport • telecom • clothes	Fleet managers • railways • airlines	All risks for a period of time determined and negotiated between producer and fleet manager		none
	Owner is producer, fleet and maintenance manager System utilization (goods and services) is sold • photocopies	Manufacturers • Aia-Geveert AG • Xerox	All risks for an unlimited period of time Producer is also fleet manager		none

SOURCE: Börlin and Stahel (1987).

TABLE 3b How Selling the Utilization of Products Rather than the Products Themselves Affects Product Design and Waste Reduction Strategies

	Alternative Utilization and Commercialization Opportunities	Impact on Product Durability	Liability for Waste and Costs of Waste Elimination	Waste Reduction Strategies
SALE	Owner is user Durable product is sold (cash or credit) • cars • white goods • clothes • computers		Costs of eliminating dispersed waste are paid by the community (local taxes)	Re-use
RENTAL	Owner is fleet manager Product is rented • cars System is rented • apartment • hotel room Service is rented • taxi	Fleet manager seeks most advantageous cost-performance ratio, which also involves speculation on tax depreciation rules.	(household waste)	Product-life extension of components Re-use
SELLING SYSTEM UTILIZATION	Owner is fleet and maintenance manager System utilization is sold • transport • telecom • clothes	Maintenance manager engages in preventive maintenance engineering. Goal is to minimize operation costs, including costs of maintenance and waste elimination.	Costs of eliminating concentrated waste are internalized by producer.	Repair Reconditioning, remanufacturing Product-life extension of components Re-use
	Owner is producer, fleet and maintenance manager System utilization (goods and services) is sold • photocopies	Producer engages in preventive engineering. The goal is zero maintenance. Full compatibility of products and systems using long-lived components.	Waste prevention occurs by optimizing the product-life of systems and components (re-use, reconditioning, leasing, updating use, etc.).	Long-life goods Repair Reconditioning, remanufacturing Technological upgrading Product-life extension of components Re-use

SOURCE: Börlin and Stahel (1987).

the materials for the production of (the same) new products. Product liability "for cradle and grave" is in most cases an upgrading of the work of the grave digger—a delegation of the problem to somebody else. Said another way: The marketing expressions "Our products can be recycled 100%" and "We recycle our products 100%" are light-years apart! One rejects responsibility and liability; the other accepts it.

By putting product use into the center of economic behavior (Figure 5), we discover that besides product-life extension there are a number of other commercial and technical strategies to intensify the use of goods. According to our reasoning that "the smaller the loop, the more profitable," these commercial strategies are even more promising than product-life extension! In addition, most of these strategies increase resource efficiency; that is, they greatly reduce the amount of resources needed to produce a given result (e.g., grams of resource per wash cycle in washing machines). The reason these strategies were invisible in the first loops (Figures 1 and 2) lies in our technocratic way of analyzing environmental problems.

So, if we broaden the definition of "reuse of goods" to include optimizing the use of goods, we find a multitude of new strategies that have the following elements in common:

- Considerable savings in resource consumption.
- Considerable waste prevention (reduction of waste in production, distribution, recycling, and disposal).
- A shift from a manufacturing economy to a service economy.
- A substitution of decentralized labor-intensive service workshops for centralized (global) highly mechanized production units.
- A substitution of labor for energy (and capital) (Figure 3) that is strongest in small loops (see Figure 2).

These strategies also imply changes in some of the basic economic concepts:

- The inclusion of the factor "time" in economic thinking; that is, the adoption of a dynamic nonlinear approach to solutions, including the simultaneous optimization of several factors.
- A change in the notion of economic value from "exchange value" to "utilization value."
- A change in the notion of risk from entrepreneurial to pure risk.
- A different attitude toward ownership where the status of utilization becomes dominant over that of property.
- A change in the function of the point of sale from a one-time point-of-no-return in the transfer of liability and full ownership to a point of service that includes periodic renegotiation.

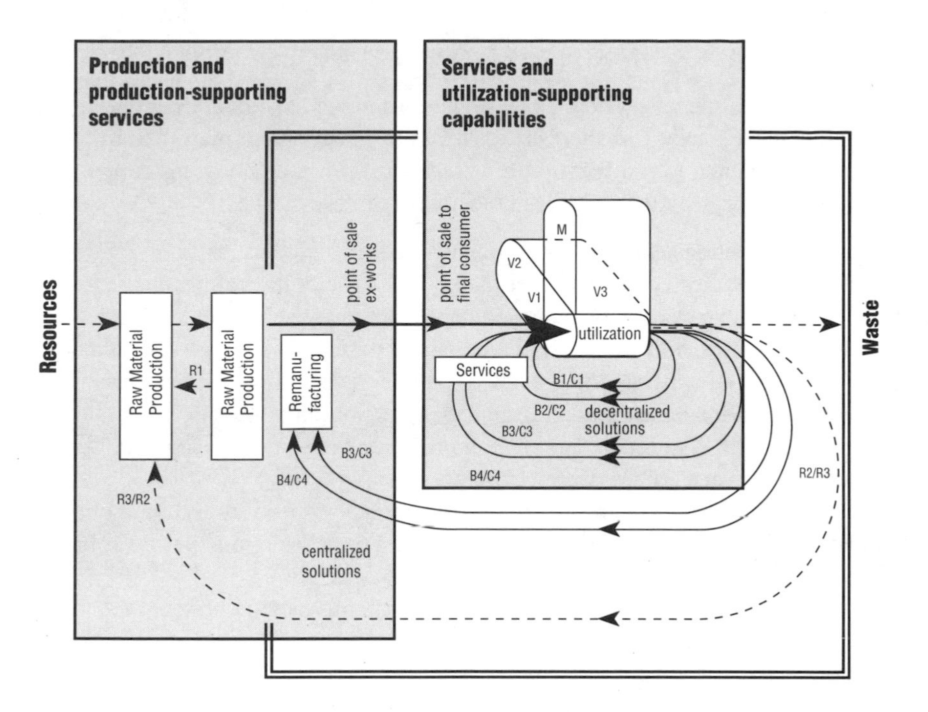

—— flows of goods	- - - - flows of secondary materials

Strategy A	Long-life goods		
Strategy B	Product-life extensions of goods	B1	Re-use
		B2	Repair
		B3	Reconditioning, remanufacturing
		B4	Technological upgrading
Strategy C	Product-life extensions of components	C1 to C4	Similar to B1 to B4
Strategy V	Commercial waste prevention strategies	V1	Selling the utilization of goods
		V2	Shared, mutual or multiple utilization
		V3	Selling quality control services instead of selling goods
Strategy M	Multifunctional products		
Strategy R	Recycling materials		

FIGURE 5 Waste prevention strategies of a utilization-oriented service economy. SOURCE: Stahel (1991).

MORE CHANGES AND OPPORTUNITIES AHEAD

If we focus technical development on optimizing use rather than optimizing production, we realize that there are vast "white" areas on our map. To shift from a product-oriented to a utilization-oriented economy, the following competitive and technological strategies, among others, become key issues:

- "Prevention" engineering (trying to construct systems, products, and components so that they could last indefinitely, with little or no maintenance).
- Adaptable system design (modular design that enables later system adaptations due to changes in technologies or user requirements through the technological upgrading of existing goods rather than their substitution for new ones).
- Risk management and consumer satisfaction on all levels, especially the system level (it is normally more economical and effective to improve system performance instead of product performance).
- Fault-tolerant system design (in an emergency, the system should give adequate warning but enable continued use of its basic functions before it breaks down).
- Self-protecting system design (long-life goods need to protect themselves and the environment against abuse by the user, such as rpm governors in car engine design to prevent speeding and engine damage).
- Technical standardization of components (the marketing strategy of redesigning every product from scratch to ensure sales of spares by original-equipment manufacturers is an economic and ecological killer in any long-life approach).
- User standardization of man-machine interfaces (for example, the standardized flight deck of modern airliners, imposed by the aircraft leasing companies, has shown the immense advantage of this approach with regard to crew cost and maintenance reliability and cost, with savings of $250,000 to $400,000 per aircraft).
- Self-curing spares (a self-explanatory concept that enables products and systems to maintain their integrity under stress, and an area where Japanese innovation appears to be far ahead of the pack).
- In-situ monitoring of system parameters and the transmission of results to a central supervisor (as used today in aircraft engine operation, e.g., by Lufthansa, and formula-one racing cars, such systems are an important part of the "self-protection" capabilities of long-lived goods).
- Training operation and maintenance engineers to the highest degree and giving them the tools to gain access to the total system knowledge through expert systems. This should, in fact, be the first priority on this list. It also implies another decisive shift in the value system of society in general and engineering in particular. Are we ready for this?

If we put all this in a cube and shake it well, we get a new definition of technical quality as system functioning over long periods of time (Figure 4). It will be hard to find a better product in the marketplace, even in 100 years' time, once a manufacturer has established itself as the "leader of quality" according to this new definition. A successful system is much harder to compete against than a successful product.

Examples of this new quality are as follows:

- Lighthouses and guidance systems that greatly increase the safety of shipping, without having to do with the design of ships themselves.
- Laundromats that allow a high-intensity use of leased long-life washing machines, resulting in a resource saving by a factor of 40 per wash cycle over typical home systems.
- Lightweight aircraft tractors that lift the front wheel of an aircraft and pull it to the beginning of the runway, resulting in a saving of 10 percent for aircraft fuel.
- Diamond—high occupancy vehicle—lanes that favor car and van pooling.
- Computers that can be upgraded by changing a component or module instead of trading in the whole computer.
- Multifunctional products, such as a laser scanner-fax-printer-copier, that result in resource savings per page, as well as in savings in (standby) energy, of approximately 75 percent.

Governments concerned for the future of the planet can take action in many ways that do not increase the budget deficit:

- Government should spend R&D money not exclusively on technical innovations, but increasingly on commercial and marketing innovations. Most of the V-strategies (commercial innovations) shown in Figure 5 could be applied today instantly without any new technology!
- Government as one of the biggest national waste producers (everything that is bought goes finally to waste) should build up a "de-curement" office at least as large as its procurement office. In the case of product reuse, one person's waste is often another person's dream! However, this other person may be at the other side of the globe and not even know the product exists, let alone that it is thrown away in working order and in large quantities.

TO SUMMARIZE

The reuse of goods, through long-life product design, product-life extension services, and strategies intensifying product use, is economically and ecologically superior to the recycling of materials. Long-life goods, product-life extension, and a more intensive use of products are part of a technoeconomic strategy within a utilization-focused service economy, which

- Is sustainable and environment-friendly.

- Maintains existing wealth and welfare.
- Is modern and economically feasible.
- Promotes technical progress and self-responsibility by economic actors.
- Reduces the speed of the flow of resources through the economy.
- Incorporates the idea of a higher resource efficiency.

The last two of these items are missing in most current "green" discussion, which focuses primarily on recycling.

The introduction of these strategies into the economy demands a change in the mind-set of corporations and government. However, these strategies are themselves long-term and are here to stay, once established. For strategic reasons, dynamic companies should therefore try to be the first in their field of activity to change.

REFERENCES

Börlin, Max, and Walter R. Stahel. 1987. La stratégie économique de la durabilité: Die wirtschaftliche Strategie der Dauerhaftigkeit. Société de Banque Suisse/ Swiss Bank Corporation, Basel. Cahier/Heft SBS no. 32, Nov. 87.

Stahel, Walter R. 1984. The product-life factor. In An Inquiry into the Nature of Sustainable Societies: The Role of the Private Sector, Susan Grinton Orr, ed. The Woodlands, Tex.: Houston Advanced Research Center.

Stahel, Walter R. 1986. Product-life as a variable: The notion of utilization. Science and Public Policy 13(4)(August):196–203.

Stahel, Walter R. 1991. Langlebigkeit und Materialrecycling - Strategien zur Vermeidung von Abfällen im Bereich de Produke. Essen: Vulkan Verlag.

Stahel, Walter, and Geneviéve Reday. 1976/1980. Jobs for Tomorrow: The Potential for Substituting Manpower for Energy. Report to the Commission of the EC. New York: Vantage Press.

The Greening of Industrial Ecosystems. 1994.
Pp. 191–200. Washington, DC:
National Academy Press.

Zero-Loss Environmental Accounting Systems

REBECCA TODD

Few people realize that existing accounting systems are a critical barrier to internalization of environmental costs and considerations by the modern firm. Simply put, you can't manage what you never see—and, with today's managerial accounting systems, managers don't see most environmental costs. All persons have a stake in the health of the environment and the conservation of scarce resources. However, managers of firms uniquely possess the necessary authority to decide which products to manufacture and which processes to employ to control how efficiently productive resources are used and, thus, to improve and preserve the quality of the environment. The information set that supports such managerial decision making derives ultimately from data collected by development and production engineers within a firm. Although accounting systems are frequently the conduits for some or all of this information, in a real sense, managers and engineers are the ultimate fiduciaries of the environment. The reason is that only the engineers know what the decision-choice set looks like and only the managers can choose.

Wastes released to the environment represent inefficient use of costly scarce resources as well as potential liability to the firm. Therefore, the reduction and elimination of wastes will, in the long run, increase firm profitability as it reduces imbedded risk. It is thus in the firm's interest to minimize wastes. Our objective must be the development and implementation on a routine basis of a *zero-loss* environmental information accounting, control, and accountability system. This system will ultimately record and monitor the flow and disposition of all inputs, including obvious waste disposal, recycling, and reprocessing. However, to approach zero information loss and, ultimately, zero environmental loss, systems must capture currently hidden and unaccountable costs, for example, product-related legal expenses, regulatory costs, public relations expenses, and the opportunity costs of clean technologies not adopted.

Unfortunately, severe institutional impediments serve to hinder managers who would otherwise pursue optimal environmental waste reduction strategies. First, traditional accounting systems were not designed to capture much of the engineering and accounting data required for environmental decision making. Second, the data that are collected and processed are almost always aggregated in such a fashion as to lose their environmental (as well as managerial control) information content. Third, line managers are rarely made responsible for environmental costs. As managers' compensation is routinely based at least partially on profitability, including reduction of controllable costs, managers have strong disincentives to seek "full environmental costing" methods that would cause additional costs to enter their control domains. Simply put, managers can't act if they don't have adequate information, and they won't act voluntarily if they will bring harm to themselves by doing so. Thus, our ultimate goals must be (1) to develop techniques for removing or minimizing the most important institutional barriers managers face in their role as environmental stewards, and (2) to provide incentives to motivate the aggressive and creative development of solutions to promote current and future waste reduction and elimination.

The specific immediate goal is to revise our traditional production accounting methods, which capture and apply to products the costs that occur in production but fail to capture any costs arising after the finished product leaves the shop floor. In this way these accounting methods disassociate the costs of waste treatment, reprocessing, and disposal (as well as regulatory costs, disposal site remediation, and contingent liabilities) from the products that produced them. We must generate feasible methods for (1) corporate accounting system capture of the costs of manufacturing wastes and effluents, and (2) the application of such costs to the individual products giving rise to them. The methods should be both specific enough to provide a basis for management decision making and sufficiently general to be applicable across a broad range of industries and national accounting systems.

In recent years, the legal, regulatory, and accounting domains have undergone significant changes that greatly enhance the likelihood that these objectives can be achieved. Indeed, certain of the changes make such change essential to the continued viability of many firms.

Application of these methods will require short-term adjustments in cost information gathering and allocation and in corporate decision making based on cost reports. However, the long-run payoffs to firms operating in the increasingly competitive global markets are higher profitability as a result of selecting the most cost-efficient products and production methods, and much lower firm risk from potential product and environmental liability. The techniques being developed will be applicable across all product costing problems, not just those associated with environmental waste.

THE INCENTIVE CLIMATE FOR ENVIRONMENTAL ACCOUNTING

Accounting may be defined as a "reconstruction" of a firm, a financial model. That is, accounting provides an image of the firm reflected from a financial mirror. In a traditional financial accounting system, only those economic events that can ultimately be defined in financial terms will be captured by the system. The information may be timely, in which case it may be highly valuable, or dated and stale, which may render the information useless. The level of aggregation at which the information is reported to decision makers is also crucial. Details are lost as information is collected into pools. As only a limited subset of all economic data available for a firm is collected, the resulting aggregated information is likely to be incomplete or flawed, depending on which economic decisions need to be made and the information necessary in a given case to support the decision.

Traditionally, accountants have held that a fundamentally important trait of accounting information to be used for economic decision making is that it be unbiased, or neutral. This property means that accountants should be detached and impartial observers of the operations of the firm rather than instigators or advocates of various courses of action. As a consequence, managers usually specify which decisions they desire to make, and accountants make recommendations regarding the information to be collected and summarized, and the analyses to be prepared in support of the decisions. Rarely is an accountant a graduate chemical engineer or operations management specialist. For accountants to provide enough relevant information to managers for environmental decision making, substantial support and advice will be required from the experts—the engineers and environmental scientists. It is not reasonable to expect managerial accountants to know the appropriate data to seek or even the general questions to ask. Consequently, routine mechanisms must be established for the cooperative collecting and reporting of environmentally relevant information.

Internal Accounting

Economic decision making is the domain of the managers of the firm. Most internal accounting systems, called managerial accounting systems, are capable of capturing any information that the management of a firm may regard as potentially useful. For example, it is common to find that the traditional accounting system has been extended to include such items as product volume information, qualitative personnel record data, and engineering data, as well as a boundless set of other information that managers at some time found it desirable to collect. Thus, managers can direct the firm's accountants to collect, analyze, aggregate, summarize, and report any information they desire, conditional only on the usual benefit/cost and other feasibility constraints. Most of the decisions about what information will be captured by the accounting system (other than that mandated by regulatory bodies) will be made by managers and other agencies outside of the accounting department.

Financial Accounting and Reporting

A subset of the managerial information collected is provided to outsiders, for example, investors, creditors, and others. Such information frequently includes voluntary disclosure of information that management may, for a variety of reasons, desire to make public. However, the greatest portion of publicly available firm-specific financial information is provided under the requirements of the two accounting rule-making bodies, namely, the Financial Accounting Standards Board (FASB) and the Securities and Exchange Commission (SEC). The FASB is the private accounting rule-making organization that provides guidelines for the public accounting profession. The SEC has the mandate of the United States Congress to oversee disclosure of financial accounting information for use in the equity and debt securities markets.

For a variety of reasons, including the pressure of competition and potential information overload by users, public disclosure of financial information is necessarily in a highly aggregated form. The information is designed to address questions of the profitability, riskiness, and viability of a firm, taken as a whole, rather than individual product profitability. Thus, the financial accounting rules do not require the collection and summarization of the finely detailed product cost information necessary to provide a basis for managers' environmental cost-reduction decisions.

Regulatory Oversight Agencies

Regulatory agencies, such as the U.S. Environmental Protection Agency (EPA), are concerned with the quality of the environment and waste minimization to enhance that quality, rather than issues of profitability. The EPA seeks, through the use of persuasion, negotiation, and regulatory tools, to measure and reduce waste emissions and effluents as well as to monitor remediation of past and current waste disposal sites. Thus, the regulatory focus is largely on the end-of-pipeline emissions and substantially less on the individual products and processes that produce the wastes, the ultimate point of control for managers. That is, the EPA and similar agencies are not in the business of micromanaging the product and process decisions of firms, but rather define the constraints under which such decisions will be made.

Managers

It is sometimes assumed that in business, as in other endeavors, more information is better than less. If so, then we should expect to find that managers routinely request and receive finely detailed cost accounting data, including breakdowns by product line of "overhead," or pooled, costs that are not direct costs of manufacturing. However, detailed breakdowns of manufacturing total cost are relatively rare. Common reasons given for the lack of such detailed cost information include difficulties in attributing so-called joint costs, costs that are shared by a variety of product lines, for example, heating costs for a large manufacturing facility; the expense of tracking and collecting numerous such costs on an individual product line basis; and, in some cases, the technical difficulty of measuring product-specific material flows and emissions. Another reason, however, relates to responsibility for costs and the use of such accountable costs in management compensation contracts as well as in cost-based product pricing formulas.

We will first consider the issue of detailed costs and management compensation. If manufacturing costs can be individually attributed to products, then managers may be held responsible for controlling the costs, and such controllable costs may well be used in profitability calculations used in management incentive compensation formulas. Thus, managers may have an incentive not to request and use the most detailed information that could be made available to them. The unrequested information may well include costs of environmental wastes, recycling and reprocessing, site remediation, and contingent liabilities.

The second problem, cost-based market pricing formulas for products, leads to the possibility that at least some of the firm's products may be found to be unprofitable to produce if all of the costs associated with their manufacture are properly attributed to the product. This would seem a simple problem with a straightforward solution: just discontinue the product. Such products may, however, account for a relatively large portion of the revenues of the firm and their discontinuance would mean loss of "market power" for the firm as well as prestige for the managers of products.

Another difficulty that managers of firms face, particularly senior management, is the confidentiality of highly sensitive information, especially that related to potentially costly contingent liabilities and regulatory constraints. That is to say, managers would have an incentive to discourage broad dissemination, even within the firm, of information deemed to have a potential negative effect on the fortunes of the firm and, consequently, the managers. In this case, the perceived risks to the firm and the managers would outweigh the possible long-term benefits from accumulating the information and attempting to control the costs. Thus, managers who may, in general, strongly support efforts to improve the quality of the environment may find that the short-term costs to themselves and their firms may outweigh the benefits.

It is thus apparent that firms, the environmental regulatory authorities, and accounting rule-making bodies have a number of incentives not to pursue highly refined cost accounting and control techniques for environmental wastes. The following section discusses in more detail what such systems might look like, and how they might best be established.

ACCOUNTING INFORMATION AND CONTROL SYSTEM COMPONENTS

Environmental accounting often means different things to different writers on the subject. I will define environmental accounting for purposes of this discussion as the incorporation and treatment of "environment"-related information in internal accounting (i.e., management accounting) systems.

The role of management accounting systems is threefold: (1) to direct managerial attention to problem areas, (2) to provide informational support for managerial decisions, and (3) in decentralized organizations, to promote the harmonization of divisional goals with corporate goals through performance measurement and incentive (management control) mechanisms. The major components of a managerial information system that includes cost accounting data collection are shown schematically in Figure 1. The first two components, standard setting and

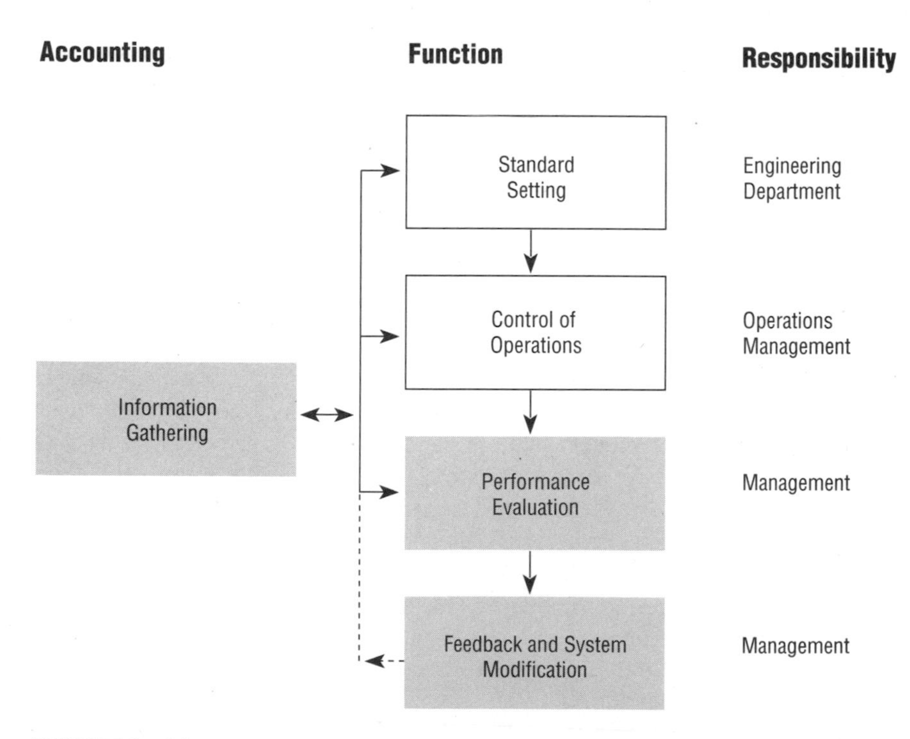

FIGURE 1 Management information system (including control system components).

control of operations, are primarily the province of the engineering departments and operations management. Standards are based on engineering process designs, actual operational data, forecasts of demand and sales of products, existing inventories, capacity constraints, and other such factors. Control of operations is an ongoing function that includes periodic monitoring of production units, yields, and quality of outputs. Both of these components are commonly based on volumes, weights, and other physical measures of productivity. Information is usually gathered in some form at every stage of production (as well as in other activities of the firm, such as sales).

At the end of some regular period of time, the information gathered is aggregated and summarized into a variety of performance evaluation reports. These reports include, but are not limited to, financial reports such as income statements, balance sheets, and cash flow information for internal as well as external dissemination, detailed cost accounting schedules for production and sales during the period, and comparisons of actual production numbers with operating budgets and forecasts. The various performance evaluation reports are then used for feedback and system modification, to control operations, improve profitability, reduce risk, take advantage of business opportunities, and reward managers who are "successful" (as measured by the performance evaluation reports).

Feedback and modification are the ultimate objectives of the management information and control systems. The success of the system depends on (1) the quality of information produced by the system, and (2) the timeliness of the information received by those in a position to modify the system.

Traditional Cost Accounting Systems

A traditional cost accounting system is shown in Figure 2. The system assumes that the firm produces two products, Widget *A* and Widget *B*, both of which use direct labor and materials. By direct labor and materials, we mean those factors that not only are associated with the production of the product, but also become a part of the product, such as raw materials, or are required to transform the raw materials, such as direct labor.

All other costs of production, including supervisory salaries, occupancy costs of manufacturing facilities, janitorial services, utilities, property taxes, materials handling costs, and disposal of wastes, including "environmental" wastes, and a host of other costs, are accumulated into "pools" of costs, usually termed "overhead" costs. Such costs are then allocated to individual products based on some systematic and rational cost allocation scheme, for example, budgeted labor hours per unit of product.

It is readily apparent that there are several problems that can have a profound effect on economic decision-making for individual products. First, important information is lost in the process of aggregation. For example, the amount of a certain waste material produced by Widget *A* will not be specifically captured for

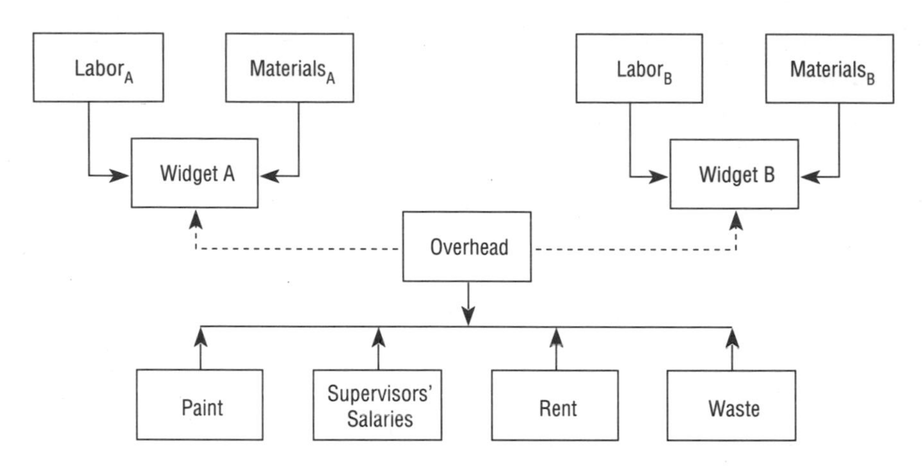

FIGURE 2 Traditional cost accounting system.

control purposes but will be added in with all other costs in the overhead pool. Second, unless environmental costs are produced by individual products at the same rate as the allocation basis (e.g., labor hours), then a misallocation of the environmental costs will occur across products. Although the logic behind the evolution of the system of aggregation of overhead in large pools (which are then allocated to individual products or product lines) may seem obscure, the explanation is straightforward. Historically, accounting data collection, aggregation, and summarization, before the wide use of computers, was a labor-intensive and extremely costly activity. Companies therefore minimized some costs, that is, accounting costs, by simplifying the accounting process at the sacrifice of more detailed cost information, relying on the budgetary process to identify costs that were "excessive."

Third, in the absence of accountability, no individual manager has an incentive to reduce or eliminate the cost. In the language of finance, each individual manager has a potential "put" to the firm as a whole for each cost that can be assigned to an overhead pool rather than treated as a direct cost. The value of the put is a function of the manager's ability to minimize the firm's consumption of the allocation base.

Let us consider an extreme example, illustrated in Figure 3. Assume that only Widget *B* produces a particular toxic chemical waste, Toxic Waste$_B$, and that the cost of disposal for Toxic Waste$_B$ is a proportionately large cost of manufacturing. In addition, we will assume for simplicity that the usual overhead allocation basis is labor hours. Under the traditional overhead cost accounting system, Widget *A* will be allocated a portion of the cost of disposal of Toxic Waste$_B$, although the manufacturing process for *A* did not require the use of the chemical. Moreover, if *A* requires substantially more labor hours than *B*, *A* will effectively subsidize the

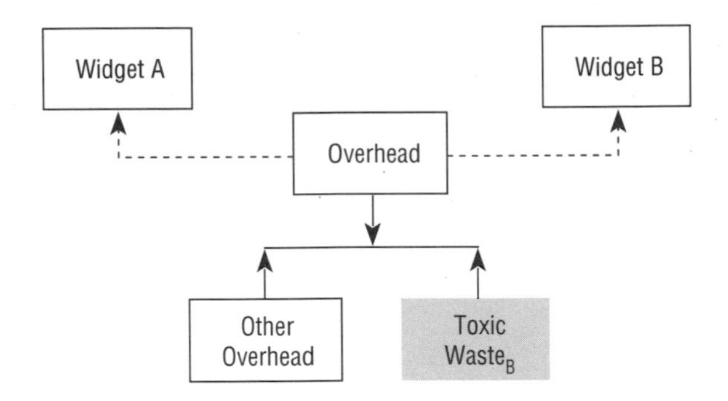

FIGURE 3 Misallocation of environmental costs under traditional cost accounting system.

production of *B* to the extent of most of the cost of the Toxic Waste$_B$. Observe that under this traditional system, the production manager for Widget *A* can reduce the costs allocated to *A*'s production only by reducing *A*'s consumption of labor hours relative to those of *B*. Clearly, the production manager of *B* has little incentive either to reduce the use of Toxic Waste$_B$ or to search for and invest in improved production technologies that will eliminate the use of the chemical.

The research question for the reduction and elimination of environmental wastes thus revolves about the problem of the assignment of the costs of such wastes to the individual product managers who are in a position to control and ultimately eliminate the wastes. Assignment of the costs requires that all costs be identified with individual products to the extent possible.

An Environmentally "Enlightened" Cost Accounting System

Figure 4 modifies the traditional accounting system (Figure 2) to indicate how the accounting and "accountability" system needs to be changed to provide both the opportunity and the incentives to the manager of Widget *B* to reduce toxic waste. If the cost of the chemical is removed from the overhead pool and applied directly to Widget *B*, just as Labor$_B$ and Materials$_B$ are applied, then the manager of *B* will have an immediate incentive to focus attention on the control of the cost. This system makes it possible to (1) ascertain the relevant controllable costs, (2) determine the operating risks (e.g., contingent liabilities resulting from the production of Toxic Waste$_B$), and (3) seek other, less costly means of production for the product or contemplate its elimination from the product line.

Perhaps the most difficult issue to resolve is that of manager *B*'s incentive compensation contract, if it is based at least partially on the profitability of *B*. Manager *B* will clearly suffer loss of wealth if the cost structure is arbitrarily changed. Thus, an essential part of the transition to a new accountability system is

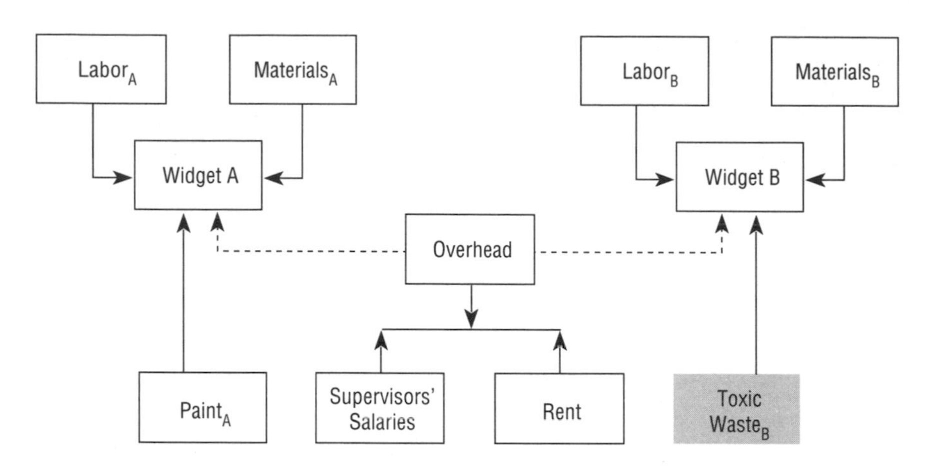

FIGURE 4 Environmentally "enlightened" cost accounting system.

to restructure the incentive compensation system to encourage the ultimate objectives of waste minimization and elimination.

Numerous possibilities exist for accomplishing this goal. Two such schemes would be (1) gradual phasing in of the new system over a period of years, and (2) basing incentive compensation on the direct cost structure for manager *A*. In addition, incentive compensation might include the direct cost structure plus a "bonus" for reduction of Toxic $Waste_B$, for manager *B*, with a target elimination horizon of, for example, five years for the chemical, over which time the full disposal costs will gradually be entered into the direct cost structure.

Modification of traditional accounting systems will require establishing hierarchies of all input and environmental costs based on difficulty of achievement, developing taxonomies of product-costing strategies for capturing and assigning all such costs to a responsible manager, and setting target horizons for reduction of environmental effluents. Nearly all the information required, including clean technology alternatives, will be provided by engineers. The system will require enormous collaborative efforts among engineers, accountants, and managers and potentially the development of new material measuring and tracking technologies as well. The potential payoff to society as well as the firm, however, is incalculable.

The Greening of Industrial Ecosystems. 1994.
Pp. 201–207. Washington, DC:
National Academy Press.

Implications of Industrial Ecology for Firms

PATRICIA S. DILLON

Environmental management in the 1970s focused on pollution control, that is, end-of-pipe treatment. The 1980s saw a shift toward waste reduction or pollution prevention. From a business and environmental perspective, treating wastes from manufacturing processes was no longer seen as the most effective or efficient solution. In the 1990s it is becoming clear that even pollution prevention as it is traditionally viewed is shortsighted. It is no longer enough just to prevent pollution from manufacturing processes. We are beginning to see instead a shift in industry and government toward a broader concept of pollution prevention—beyond the manufacturing process—to encompass environmental concerns and pollution prevention throughout the life cycle of a product, from acquisition of raw material to ultimate disposal (Figure 1).

In leading firms, the concept of product life cycle management is being formalized in Design for Environment (DFE) and product stewardship programs, aimed at the design of more environmentally compatible products and the extension of producer responsibility beyond the manufacturing plant to influence product use, recycling, and disposal by customers. For example, as discussed by Klimisch in this volume, automobile manufacturers are redesigning vehicles to facilitate the reuse, recycling, and reclamation of materials and components, particularly nonmetallic components. As part of this effort, they are working with suppliers, car dismantlers, and shredder operators to create the necessary infrastructure to recover, recycle, and market vehicle components and materials after the automobile's useful life.

By focusing on the entire life cycle of a product rather than discrete sources of pollution, companies are clearly moving in the right direction. In addition,

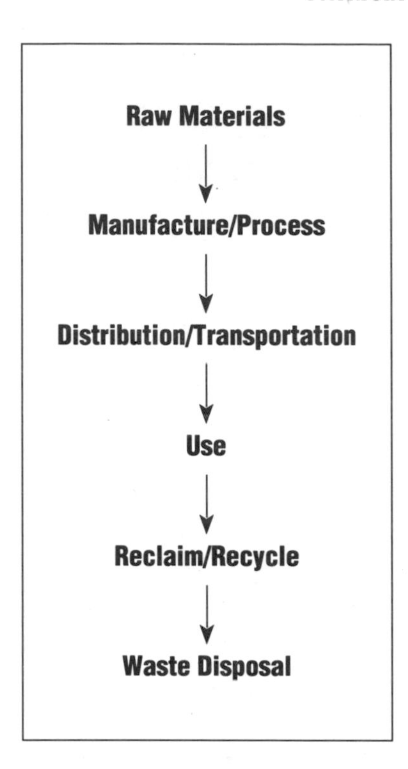

FIGURE 1 Product life cycle.

attitudes in progressive firms are evolving to the point where environmental issues are no longer viewed as problems to be defended against or prevented. Instead, the environment presents opportunities to transform management practices to make environmentally superior products and services and gain competitive advantage (Booz, Allen and Hamilton, 1991).

Although firms are making progress on the environmental front, current efforts will not significantly improve the environmental characteristics of the current industrial ecology. If the ultimate goal is a closed industrial ecosystem that consumes limited resources and produces limited wastes (Frosch and Gallopoulos, 1989), then industry and society cannot simply minimize product impact on the environment. Attention must also focus on optimizing material flows and integrating the life cycles of diverse products, so that waste streams from one activity become the raw material for another.

Such changes will require radical transformation in industrial practices and relationships. In most cases, technology probably will not be the limiting factor, a lesson learned in recent years as companies have adopted pollution prevention practices. Rather, cultural and organizational changes within industry (as well as changing the behavior of consumers and government agencies) will most likely present greater obstacles.

As a point of departure, this paper first examines the implementation of DFE and product stewardship programs in firms because they share a systems-oriented approach consistent with industrial ecology thinking. They focus on reducing environmental risks over the life cycle of products rather than merely reducing pollution at the manufacturing site. Evaluating existing efforts within firms should provide some insights into how to proceed in the future. This paper then examines where current industrial efforts fall short of meeting environmental goals and what additional steps might be needed to move closer toward a closed industrial ecosystem.

The remarks in this paper are based on a two-year study at Tufts University Center for Environment that examined the scope, design, and implementation of corporate product responsibility programs, such as Design for Environment and product stewardship.

MAKING THE CONCEPT OPERATIONAL

Companies face numerous challenges as they try to implement an "environmental program" that affects virtually all aspects of their business. To capture environmental considerations throughout the product life cycle, the program must reach out and link many corporate functions, which may not have been traditionally involved in environmental issues. For example, the program must influence the product development process at a point where the selection of alternative materials is feasible and not cost prohibitive; extend to purchasing in order to shape the activities of suppliers; incorporate input from enviromental, legal, and regulatory experts; solicit the input of marketing to monitor customer requirements and expectations; and tap into technical assistance departments to provide environmental services to customers.

Although company product responsibility programs are still in the developmental stages and are primarily the initiative of large, progressive firms, some common and apparently successful features of their programs emerge.

- **Commitment of senior management**. Clear and prominent leadership from senior management is necessary to achieve cultural, organizational, and procedural changes such as those outlined below. For example, the Environmentally Conscious Products initiative at International Business Machines (IBM) Corporation was the direct result of a senior vice president's efforts to address European customers' concerns about product disposal and the shortage of landfill space. A worldwide task force comprising more than 50 representatives from diverse functional areas within IBM was formed to address the issue. The emphasis of the task force shifted quickly from its initial focus on disposal to the entire product life cycle, as the task force realized that any disposal improvements would come only from product design changes. One important outcome of this task force was IBM's guidelines for environmentally conscious products.

• **Development of companywide strategies and guidelines.** Corporate guidance is needed to ensure that product responsibility is carried out appropriately and consistently within a company. With input from various functional groups, environmental and technical staff often provide overall direction and oversight of the program, including delineation of environmental goals and priorities (such as design for recyclability or reduction in toxic emissions) and an overall strategy for accomplishing these goals. Companies are developing guidelines for program implementation (for example, procedures for integrating environmental considerations into product design processes) and corporate policies and standards for priority concerns.

3M Corporation's Product Responsibility Program encourages business units to think holistically—from cradle-to-grave—about their products and to develop plans to address environmental issues that arise during product manufacture, use, and disposal. To help business units identify problem areas and opportunities to improve the environmental performance of products, corporate staff developed two tools, a self-evaluation survey and a life cycle model for product responsibility. The life cycle model outlines activities and issues for business units to consider, such as minimizing the use of toxics in products or processes, incorporating reusable and recyclable materials in products, and identifying opportunities for product and package recycling by customers. In order to avoid potential misuse of environmental claims, 3M also issued a corporate policy covering environmental marketing and established a formal process for the review and approval of environmental claims, symbols, and slogans.

• **Creation of multidisciplinary networks.** Since product responsibility infiltrates so many aspects of a business, it is important to develop networks or teams of individuals with different training and perspectives on the business to solicit the input of appropriate functions in decision-making processes and to provide a conduit for information exchange. Multidisciplinary networks can be used in the formulation of strategy and goals for the overall product responsibility program; in responding to environmental challenges facing individual product lines; and in dissemination of relevant regulatory, technical, and internal management developments. Depending on the issue at hand, involvement of personnel from research and development, purchasing, technical service, legal and regulatory, marketing and sales, and health and safety groups may be appropriate.

• **Harmonization of product responsibility program and goals with other company practices and goals.** To ensure that environmental issues are considered along with traditional criteria such as cost, quality, and performance, companies are integrating environmental product responsibility objectives into existing and familiar decision-making processes and tools such as review phases for new products and expert systems. For example, the original premise of DFE among companies such as AT&T and automobile manufacturers was to incorporate environmental principles into their "Design for X" (DFX) process, where "X" represents any product characteristic such as reliability or manufacturability that the company wants to maximize in its product design. At Xerox Corporation total quality management (TQM) practices provide the foundation for its product improvements for the environment.

The involvement of the manufacturing and development organizations, as well as functions such as marketing, is critical to the integration of the concept into the fabric of the business. For example, involvement of product managers and design engineers is necessary to make the concept and practice part of the product development process; sales, marketing, and technical service staffs must also embrace the concept of environmental product responsibility if the company hopes to deliver its expertise and services to customers. In addition, the application of environmental product responsibility principles and guidelines by product groups recognizes the relative autonomy of product lines and provides the flexibility for product groups to apply their expertise to the product's technical and performance specifications, as well as its uses and markets, to identify environmental issues and solutions and to fine-tune the environmental product responsibility program to their needs.

• **Establishment of a systematic and iterative process**. Carrying out a comprehensive analysis of the product life cycle and identifying possible measures to reduce environmental risks involves consideration of environmental, regulatory, legal, technical, and market issues. The broad scope of this task necessitates a deliberative, systematic evaluation process to ensure that the viewpoints and data provided by members of the multidisciplinary group are incorporated into business decisions. In addition, since environmental regulations, societal values, and customer requirements are forever changing, products must change over time. Therefore, the program must provide for periodic evaluation of products and opportunities for continuous environmental improvements.

FUTURE CHALLENGES

As firms adjust their management practices and decision-making processes to meet the challenges of managing product life cycles rather than pollution sources, they begin to create an atmosphere in which movement toward a closed industrial ecosystem is possible. However, current efforts of U.S. firms tend to be ad hoc, even when the goals of the product responsibility program are far reaching. Companies' efforts tend to address single issues (for example, reduction of toxic emissions from manufacturing operations, increasing recycled content in products) or particular stages in a product's life cycle (for example, disposal). Integration and cooperation among firms, which is necessary because the activities of individual firms do not span the entire product life cycle, is also limited at the present time. In addition, little attention is being paid to some critical issues such as reducing raw materials consumption and extending the useful life of products. A critical question, therefore, is how to move beyond current efforts toward a more comprehensive product life cycle management and, ultimately, a closed industrial ecosystem?

Company initiatives are largely influenced by external signals that have economic implications for the firm (for example, current and pending regulation and issues on the agenda of the public.) While customer requirements and demand also play a role, customers themselves are frequently driven by these same external signals. Several examples help to illustrate this point.

- As an extension of the company's product stewardship philosophy, Dow Chemical in 1990 formed a new business group, Advanced Cleaning Systems (ACS). Unlike previous product stewardship efforts at Dow, ACS was designed principally as a business strategy: its goal, to maintain and increase sales and profitability in the company's chlorinated solvents business, which is threatened by increased regulation of ozone-depleting chemicals and federal and state programs to control toxic air pollutants. To accomplish its goal, ACS is developing alternative chemicals and processes as well as a range of customer services, including improved containers and process controls, aimed at reducing the environmental impact of Dow products from their arrival at the customer's facility through processing and waste handling.
- In the automotive and electronic industries, design for disassembly and recycling initiatives are largely influenced by international concerns such as rising disposal costs and legislation in Europe, specifically German postconsumer "takeback" legislation, and the possibility of U.S. adoption of such measures. Such legislation generally requires manufacturers to take their products back after consumers are through with them, and recycle or dispose of them properly.
- While the United States does not yet have national packaging regulations, individual states and nongovernmental organizations such as the Coalition of Northeastern Governors (CONEG) are responding to solid waste disposal issues by developing model legislation dealing with heavy metals in packaging and packaging waste reduction. These efforts have prompted actions by companies to redesign their packages and reduce packaging volume.

Improving the environmental characteristics of entire industrial ecosystems will not be achieved by putting out fires or reacting to the public policy issues of the moment. The public policy agenda does not always reflect the most important environmental priorities and risks, long-term issues, or cross-media environmental concerns. Rather, it often reflects short-term priorities, perceived risks, and politically "hot" topics, and addresses single environmental media issues or contaminants (U.S. Environmental Protection Agency [EPA], 1990, 1992). In addition, since company efforts focus on activities in which they derive economic benefits, realization of industrial ecology goals will necessitate getting less profitable ventures on the corporate agenda: for example, recycling of materials and wastes that have little value; providing services to small, dispersed customers; and identifying and addressing issues for which there are minimal or nonexistent regulatory, public, or customer pressures.

Incentives for firms can be created by national environmental policies and legislation (such as the German take-back legislation, Japan's Law for Promotion of Utilization of Recyclable Resources, and the Dutch National Environmental Policy Plan) or voluntary programs (such as EPA's Industrial Toxics and Green Lights programs) that encourage comprehensive, long-term solutions to environmental problems through the development of more environmentally sensitive products and the creation of industrial infrastructures and partnerships. For example, the German take-back legislation requiring the recycling of all packaging is forcing the development of a private recycling infrastructure ("Green Dot" program) and the redesign of packaging (Cairncross, 1992). Existing regulations must also be examined to eliminate unnecessary or inappropriate barriers, such as hazardous waste regulations that might discourage the take-back of chemical products by suppliers or antitrust regulations that limit industrial cooperation.

Setting future environmental priorities and goals in both the private and the public sector also will be facilitated by the further development of analytical methods, such as life cycle analysis, that better capture the true and total life cycle impacts of products and processes on the environment. In the near term, the challenge for industry and government is to set priorities and make decisions that will hold up in the face of advancements in analytic methods such as life cycle analysis and will not become costly mistakes to the company and the environment in the future.

REFERENCES

Booz, Allen and Hamilton. 1991. Corporate Environmental Management: An Executive Survey. Bethesda, Md.

Cairncross, Frances. 1992. How Europe's companies reposition to recycle. Harvard Business Review 70(2):34–45.

Frosch, Robert A., and Nicholas E. Gallopoulos. 1989. Strategies for manufacturing. Scientific American 261(3):144–152.

U.S. Environmental Protection Agency, Science Advisory Board. 1990. Reducing Risk: Setting Priorities and Strategies for Environmental Protection. SAB-EC-90-021, Washington, D.C.

U.S. Environmental Protection Agency. 1992. Safeguarding the Future: Credible Science, Credible Decisions. EPA 600/9-91/050, Washington, D.C.

The Greening of Industrial Ecosystems. 1994.
Pp. 208–213. Washington, DC:
National Academy Press.

Design for Environment: An R&D Manager's Perspective

ROBERT C. PFAHL, JR.

The Design for Environment (DFE) concept and its implementation offer manufacturing companies the opportunity to achieve world-class economic performance by producing world-class products, which increasingly means products that are environmentally acceptable throughout their life cycles. Based on the principles of industrial ecology, DFE is a means to achieve environmentally conscious designs, which are a necessary step toward sustainable economic development. Even though the American electronics industry is a secondary sector—an industrial sector concerned with converting processed materials into manufactured products—and thus a relatively "clean" industry with minimal waste (Ayres, 1992), DFE can play a significant role in decisions related to manufacturing.

With more than 2.4 million employees, electronics is the largest industrial sector in the United States. Some firms produce thousands of different products, each with hundreds of components, most of which are procured from other firms in the industry. Approximately 70 percent of all firms in the industry are small companies with fewer than 200 employees. Thus, the infrastructure is extremely complex, with many firms functioning as both suppliers and customers.

The heart of electronics design is the electrical and mechanical interconnection of individual components and subassemblies by fastening them to a substrate that contains electrically conductive wiring. These designs are manufactured by using mass-production processes that accommodate the assembly of a broad range of components.

Because of the rapid advances occurring continuously in electronic components—advances that reduce size and cost while increasing speed and functionality—periodically the performance of the component packaging or the interconnection substrate is exceeded and new technology (including design, materials, and manufacturing technology) must be introduced (Pfahl, 1992). Thus, unlike many other manufacturing industries in the secondary sector, the American electronics industry periodically goes through fundamental shifts in manufacturing technology.

At present, printed wiring board (PWB) technology is reaching its limits for certain applications, and many electronics companies are investigating multichip modules and other technology alternatives. These companies are motivated to practice DFE in this application, because about one-third of the cost of current PWB technology is associated with the waste that occurs in processing and manufacturing. However, they find technological solutions constrained by a myriad of local, state, and national regulations. In a command-and-control environment, regulations are designed for the existing industry and by-and-large assume static technology (indeed, by mandating certain control technologies, such regulations frequently freeze technological evolution). Unfortunately, however, in an industry that is changing rapidly and practicing DFE, regulations can serve to preclude major paradigm shifts in industrial ecology from a linear, Type I, system, which requires unlimited resources, to a recycling, Type II, system in which limited resources are used and limited wastes are produced (Allenby, 1992a; see also Richards et al., in this volume).

The following three experiences, drawn from an R&D manager's perspective in the American electronics industry, illustrate environmental issues being addressed today. From these experiences, industrial ecologists can draw conclusions about (1) the feasibility and benefits of widespread implementation of DFE; (2) today's command-and-control regulatory system, which in too many instances is hindering the establishment of proactive industrial programs; and (3) the challenges to be faced in developing public policy and regulations that will enable secondary industries with complex processes and products to change the current industrial ecology to one that is environmentally more sustainable.

THREE EXPERIENCES

The Cost Reduction Experience

Cost reduction is a major activity in most manufacturing firms. It is driven by establishing cost-reduction metrics for manufacturing engineers. Consider the case of a manufacturing engineer who sought to reduce the cost of painting sheet metal used in electronic equipment assemblies. He discovered that he could lower costs by having the sheet metal components chromated (plated with a chromium compound). Since many of the high-quality coating firms no longer provide chromate coatings, because of increased environmental regulation, the new chromate coatings were applied by a second-tier supplier. Potential quality failures were identified during accelerated reliability testing. The effort ended with the manufacturing engineer being advised that the proposed change did not provide acceptable protection and that future use of chromate coatings should be considered as a last resort because of potential future environmental costs.

In this experience the indirect impact of environmental legislation was to reduce the quality of chromate coatings to an unacceptable level, which in effect made this technology obsolete. While this may or may not be desirable, it is clear that no systems-based analysis preceded this regulatory restructuring of available technologies. There was no evaluation, for example, of the environmental impacts of the alternative technologies, or balancing of these impacts against those associated with the chromate process.

Internally, in making his decision to change the coating, the engineer gave no consideration to environmental impact in terms of either product quality or environmental risk. This approach led to wasted time, effort, and resources and, if implemented, might have led to future liabilities for the firm. The failure of the firm to properly define costs for its engineering community, and to adopt a comprehensive approach to cost definition, was the direct cause of this problem. Knowledge and application of DFE principles to this design problem would have broadened the factors the engineer considered in selecting an alternative coating to achieve his cost-reduction metrics and reduced the costs and risks of this technology decision for the firm.

The CFC Elimination Experience

In 1988 the Montreal Protocol established an international phaseout schedule for the use of chlorofluorocarbons (CFCs), which had been identified as causing a rapid depletion of ozone in the earth's stratosphere. Subsequently the 1990 Clean Air Act established even tighter regulation of CFCs in the United States. About 12 percent of the total domestic use of CFCs was as a solvent for the cleaning of electronic components and precision metal parts, using azeotropes of CFC 113. The U.S. electronics industry needed to implement a set of environmental regulations with significant technological repercussions.

However, the situation differed from prior industrial experiences. In the past when a new requirement was established for air emissions or wastewater discharges, a company's environmental or facilities organization could contract with a supplier to design and build the necessary "end-of-pipe" equipment to make the facility compliant. The CFC elimination experience was different in three important aspects:

1. There were no generic alternatives because the CFCs were used as solvents in a significant number of radically different manufacturing processes.
2. Either alternative technologies to replace the CFCs did not exist, or their efficacy had not been demonstrated.

3. The necessary adjustments had to be made by manufacturing and design engineers, not by using additional end-of-pipe controls.

Finding alternative solvents and processes required the technical expertise of the manufacturing process engineer. Thus, environmental issues had to be identified, studied, and resolved by the manufacturing organization rather than by an environmental staff organization. The environmental staff could serve as an effective facilitator, but the engineer responsible for production had to propose an alternative, and the reliability organization had to confirm its effectiveness. The impact on the organization was similar to that of introducing total quality management, in which quality is everyone's responsibility, particularly the line organization's, and not just the quality or staff organization's responsibility. The success of the resulting programs at a number of American electronics firms has been recognized by the U.S. Environmental Protection Agency (EPA) through presentation of Stratospheric Ozone Protection Awards.

In this instance, implementing an "environmental" regulation required, for the first time on any significant basis, the involvement of "line" manufacturing and design organizations. Moreover, the success of the electronics industry was achieved partially as a result of the enlightened regulatory approach: establish a goal (stop the use of CFCs), but don't have nontechnical regulatory personnel try to micromanage the process. Moreover, an important concomitant benefit of the "goal-oriented" approach was the fostering of a more cooperative, consensual effort among regulators, industry, and some environmental groups, as opposed to the adversarial relationship implicit in the command-and-control approach.

The Solder Dross Experience

Lead-tin solder alloys are used by the electronics industry to attach components onto copper traces on printed wiring boards. One method for performing this attachment is to use a wave soldering machine, which produces a fountain of molten solder. During soldering, however, some of the hot, molten solder oxidizes, producing a dross that is unusable in the process and must be skimmed from the machine. For many years the electronics industry recycled this dross to secondary smelters, thereby reducing the net demand for mined lead (Allenby, 1992a). Unexpectedly in 1991, solder dross was declared to be a hazardous waste in a letter prepared by an EPA staffer without industry input. This had the effect of stopping the ecologically beneficial lead dross recycling program because most of the secondary smelters were not licensed to handle a material defined as hazardous waste under the Resource Conservation and Recovery Act (RCRA). (RCRA involves substantial paperwork and reporting burdens and imposes substantial potential liabilities on any facility handling material defined as hazardous waste.) Managers felt trapped between being environmentally responsible or obeying the regulations. After a series of meetings, the EPA backed off somewhat from its regulatory reinterpretation, which in turn allowed some of this recycling activity to resume. It was apparent, however, that the agency personnel who had written the letter had little technological or industry background and had not considered the deleterious impact their reinterpretation would have on the environment.

This case illustrates a major and growing cost associated with the adversarial, command-and-control system that characterizes environmental regulation in the United States. Environmental regulators in the United States seldom have any significant technological experience or education. This can, as in this case, lead to decisions that superficially appear to be environmentally sound—"treat all lead solder dross as hazardous waste"—but in the practical, complex, and highly technological world, are environmentally suboptimal. Incidentally, there are some indications that the more consensual environmental management systems in Europe and Japan result in more technologically sophisticated and efficient environmental regulations (Allenby, 1992b; 1992c), which may also give firms operating in those areas some competitive advantages as well.

CONCLUSIONS FROM THE THREE EXPERIENCES

The Cost Reduction Experience illustrates how a firm that fails to integrate environmental costs and considerations into its technology decisions can create higher economic costs for itself, even in the short term. Conversely, the CFC Elimination Experience suggests that DFE, an integrative approach to environmental issues, can be employed successfully in a manufacturing organization where the regulatory structure so permits, even when the technological and environmental concerns are quite significant and apparently intractable.

Internalization of environmental considerations through DFE is, however, necessary for the firm to respond successfully. Modifying corporate culture is difficult, but fortunately the quality programs of the late 1980s have paved the way for introducing DFE. When the concept of DFE was introduced to engineers at Motorola, for example, they grasped it quickly and related it to their previous experience with Motorola's Six Sigma Quality program and CFC Elimination Program. Just as quality is the entire corporation's responsibility and the internal quality organization is responsible for facilitating and measuring the activity, environmental stewardship is the entire corporation's responsibility and the internal environmental organization is responsible for facilitating and measuring the activity. Motorola is taking advantage of the strong emphasis management has placed on quality as Motorola University develops a companywide training program on DFE.

The Solder Dross Experience demonstrates how command-and-control regulatory tools can be counterproductive in enabling industry to develop proactive recycling initiatives. The ambiguous and shifting federal and state "hazardous waste" definitions and regulations are currently, for example, a significant detriment to ecologically beneficial recycling. Moreover, the lack of technological and industry expertise on the part of environmental regulators is an increasing impediment to real environmental progress as environmental regulation moves beyond the "end-of-pipe" approach.

This is not to argue that command-and-control regulations are not necessary to establish high standards of industrial behavior to which all firms must be held. It is to say, though, that the United States is overbalanced toward command-and-control regulation and the adversarial attitudes it engenders, to the detriment of both the environment and the economy. We cannot get to sustainability by relying on command and control.

As one considers developing policies that will stimulate shifts to environmentally preferable industrial ecosystems, one must be careful to create a structure that accommodates—indeed, takes advantage of—the major paradigm changes in product and process technology that occur in industries such as the electronics industry. Policymakers must recognize, as they evaluate alternative policies that encourage DFE practices, that industries experiencing periodic changes in their product and manufacturing technology provide a unique opportunity for implementing major changes with positive environmental impact. These changes in secondary sector industries such as the electronics, automotive, and aircraft industries can have a leveraged impact on primary-sector industries (those concerned with the extraction of raw materials [Ayres, 1992]) by reducing, changing, or eliminating the demand for raw materials (Allenby, 1992a). However, the risks from an unknown and fluctuating regulatory environment at present discourage corporations from considering major environmentally preferable changes in manufacturing processes and materials.

REFERENCES

Allenby, B. R. 1992a. Industrial ecology: The materials scientist in an environmentally constrained world. MRS Bulletin 17(3)(March):46–51.

Allenby, B. R. 1992b. Trip Report—Europe, National Academy of Engineering, May.

Allenby, B. R. 1992c. Trip Report—Japan, National Academy of Engineering, May.

Ayres, R. U. 1992c. Toxic heavy metals: Materials cycle optimization. Proceedings of the National Academy of Sciences 89(February):815–820.

Pfahl, R. C., Jr. 1992. Materials in electronic manufacturing: Electronic packaging. MRS Bulletin 17(4)(April):38–41.

Education and Research Needs

The Greening of Industrial Ecosystems. 1994.
Pp. 217–227. Washington, DC:
National Academy Press.

The Two Faces of Technology: Changing Perspectives in Design for Environment

SHELDON K. FRIEDLANDER

The tension between technology and environment has become a central political and economic issue both nationally and internationally. This issue has come to a head over the last 30 or 40 years, starting with the popular ecological works of Rachel Carson and the air pollution discoveries of Arie J. Haagen-Smit.

The British author C. P. Snow (1971) wrote cogently about what he called the "two faces" of technology, benign and threatening. "All through history it has brought blessings and curses: . . . It was true when men first made primitive tools and clambered out into the open savannah: one of the earliest uses of those tools seems to have been for homicide. It was true of the discovery of agriculture: which transformed social living, but also made some sort of organized armies practicable. It was true of the first industrial revolution. Perhaps the sharpest example of this two-faced nature of technology is the effect of medicine. . . . It has reduced infantile mortality, even in the poorest countries. . . . Yet it has led us straight into the flood of population which is the greatest danger of the next fifty years."

Snow did not refer directly to the technology-environment confrontation emerging at the time he wrote these words. However, he went on to write: "The only weapon we have to oppose the bad effects of technology is technology itself. There is no other. It is only by the rational use of technology—to control and guide what technology is doing—that we can keep any hopes of a social life more desirable than our own: or in fact of a social life which is not appalling to imagine." This paper takes up this theme as it applies to the design of engineering systems for pollution prevention.

There will have to be major changes in the way we design technological systems—both engineering processes and products—for minimum pollution. This represents one of the greatest challenges currently facing the engineering profession. The process has barely begun, and only faint outlines of how and where we are going are visible. As a starting point, we ought to identify those engineering design procedures or paradigms that can best be adapted to the design of clean technologies. In some cases, existing procedures suitably modified will be directly applicable. An example taken from chemical engineering is given below. In other cases, it will be necessary to invent new procedures. This may well be true of important classes of consumer product design.

In what follows, extensive use is made of chemical engineering design paradigms to follow both the conversion of raw materials to chemical products and the simultaneous generation of pollutants. The chemical engineering approach makes use of material balances, chemical kinetics, thermodynamics, and transport processes to track the conversion of a set of reactants (raw materials) into a set of products. The power of the approach makes it applicable not only to chemical plants and refineries but also to power plants, microelectronics processing, aerospace factories, and other industries that make extensive use of chemical processes. It can be applied at several different but interacting scales (Figure 1). An example at the largest scale is the work of Ayres (1989) on "industrial metabolism," which makes extensive use of material balances in tracking the large-scale flow of metallic elements through various industry sectors and into the environment.

Macroscale analyses (such as "industrial metabolism" or the closely related "industrial ecology") are useful in developing national and international strategies for reducing material losses to the environment and planning future technological development. On a mesoscale—the individual chemical plant or petroleum refinery—the approach is used to design plants for the conversion of raw materials, such as crude petroleum, or a particular chemical feedstock, such as propylene, to desired products. Finally, at the microscale this approach is employed in the design of individual chemical reactors, employing differential balances on elemental reactor volumes. The three scales of organization are closely linked as shown in Figure 1.

While instructive, rates of flow of chemical species through industry and environment tell only part of the story. Chemical compound form and the physical and chemical properties of the mixtures may strongly affect public health and ecology. In addition, concentrations change by orders of magnitude during processing and environmental transport and transformation. The concentration has a determining influence on the economics of recovery and reuse (see Allen and Behmanesh, in this volume).

The extension of the chemical engineering paradigm to the formation and control of *undesired* chemical species—pollutants—is in some cases straightforward at the meso (manufacturing plant) and micro (chemical processes) scales. However, as discussed later in this paper, the design skills required for predicting

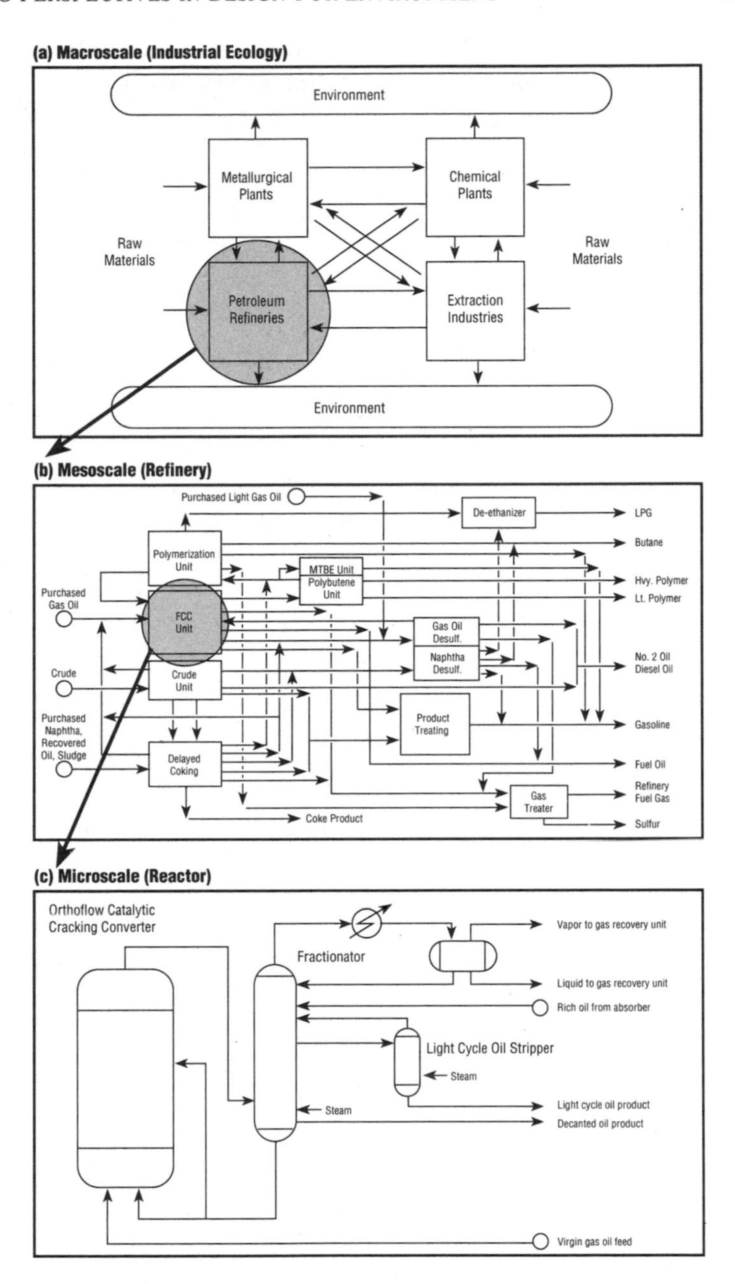

FIGURE 1 The flow and transformation of chemical substances in industry is a key feature in waste minimization or pollution prevention. Analysis of such flows can be conducted at (a) the macroscale (industrial metabolism or industrial ecology), (b) the mesoscale or individual manufacturing plant (e.g., refinery), or (c) the individual chemical reactor. Scales (b) and (c) correspond to chemical process engineering.

undesired by-products will frequently be beyond those normally applied to the high-yield conversion of raw materials to desirable products.

Changing design perspectives reflect the response of the engineering profession to a shifting national consensus toward further reducing emissions from man-made sources. The public, while still eager to share in the benefits of technology, has become much less forgiving of the associated environmental consequences. The history of this shifting national consensus is discussed in the next section, which sets the stage for the new design synthesis now under way.

HISTORICAL PERSPECTIVES: FRANKLIN AND THOREAU

The two schools of thought regarding the benefits of technology and its environmental effects can be summarized in a much simplified way as follows:

One: Technological development is fundamentally good and proceeds inexorably because it makes life better for mankind; undesirable side effects can be avoided or contained.

Two: There are limits to technological growth imposed by environmental, energy, and resource constraints; development must be slowed because of uncertainty in the side effects of technology.

The tension between these views has been reflected in the writings of some of our greatest scientists and intellectuals since colonial days. Examples are Benjamin Franklin, a leader in the development of applied science, and Henry David Thoreau, often skeptical of the benefits of technology.

Benjamin Franklin, the first great U.S. scientist, made fundamental contributions in many branches of science, especially electricity, where he introduced the concept of positive and negative charges. But he was also an applied scientist and engineer who delighted in inventing new devices to satisfy human needs.

Franklin incorporated environmental and safety concerns in the design of his inventions. He designed his stove for lower emissions and higher fuel efficiency than conventional stoves. To promote the Franklin stove, he wrote and published a pamphlet entitled "An Account of New-Invented Pennsylvania Fireplaces: Wherein their construction and manner of operation is particularly explained, their advantages above every other method of warming rooms demonstrated; and all objections that have been raised against the use of them answered and obviated."

He noted that his stove "cures most smoaky chimneys, and thereby preserves both the eyes and the furniture. . . . It prevents the fouling of chimneys; much of the lint and dust that contributes to fouling must pass through the flame where it is consumed" (Franklin, 1744). Franklin had a good appreciation of the importance of designing technological systems for low emissions.

Henry David Thoreau, who came a century later, was this country's greatest proponent of the limits-to-growth, small-is-beautiful school. Thoreau was the archetypical skeptic when it came to the benefits of technology. Indeed this was a central theme of *Walden, or, Life in the Woods*. Referring to the new telegraph cable, for example, he wrote, "We are eager to tunnel under the Atlantic and bring the old world some weeks nearer to the new; but per chance the first news that will leap through into the broad, flapping American ear will be that Princess Adelaide has the whooping cough."

In 1843 Thoreau reviewed a remarkable book by one J. A. Etzler, a German immigrant to the United States, called *The Paradise Within the Reach of All Men, Without Labor, by Powers of Nature and Machinery*: "I promise to show the means of creating a paradise within ten years, where everything desirable for human life may be had by every man in superabundance, without labor and without pay."

Etzler's idea was to develop solar "burning mirrors" to boil water and produce steam; he proposed to build huge windmills and tide mills to generate power. He foresaw large-scale desalination of water and heavy earthmoving and leveling equipment. He called for mass immigration and federal subsidies for railways, canals, and power plants. "Man is powerful but in union with many. Nothing great, for the improvement of his own condition or that of his fellow men, can ever be effected by individual enterprise."

Thoreau was impressed, in spite of himself, by the breadth of Etzler's vision and his supporting calculations. However, he rejected the basic materialism and loss of individualism expressed in Etzler's work: "The chief fault of this book is that it aims to secure the greatest degree of gross comfort and pleasure *merely*." I emphasize the word "merely"; for Thoreau, comfort and pleasure were not the greatest good. Self-sufficiency and harmony with nature were more important.

Over the last 40 years, there has been a shift in public attitude toward what might be called the Thoreau school of skepticism regarding technological development. During that period Rachel Carson, Arie Haagen-Smit, Mario Molina, Sherwood Rowland, and others have discovered (and popularized) extraordinary and unanticipated side effects of the first order resulting from chemicals released to the environment.

Understanding these effects came from the ability to puzzle out hidden relationships between the quality of the environment and the chemical releases. Scientists were able to determine the chain of events that lead from sources of air or water pollutants to their environmental consequences. A leading role has been played by scientists trained in chemical sciences.

Photochemical smog is a case in point. The origin of photochemical smog was first elucidated in the early 1950s by Haagen-Smit, a bio-organic chemist who worked in the field of taste and odor chemistry. He was asked by the Los Angeles air pollution control authorities to apply his flavor separation techniques to determine the components in the irritating air that people in Los Angeles were complaining about. From several hundred cubic feet of air (an equivalent of the amount a person breathes in one day) he collected a few ounces of condensed liquid. This liquid, mostly water, contained foul-smelling components that he identified as aldehydes, acids, and organic peroxides. Although these substances were known to be products of incomplete combustion, they were not known to be significant air pollutants. Their discovery in the condensates provided Haagen-Smit with the clues he needed to figure out the origin of the smog: organic compounds present in the air—mostly hydrocarbons—were oxidized through the combined action of oxides of nitrogen and sunlight to form the substances that produced eye irritation and plant and materials damage.

The sources of the hydrocarbons and nitrogen oxides were automobiles, power plants, and refineries. (Now we think there may be significant natural, or biogenic, contributions to the organic vapors.) But the original emissions themselves (primary pollutants) were not the main problem—*it was the reaction products that formed later in the atmosphere (secondary pollutants) that caused severe unanticipated environmental effects*. The relationships between air quality and emission sources were complex and would have been difficult to predict. Indeed this is a conspicuous characteristic of the modern age of technology/environment interactions—unanticipated side effects resulting from complex chemical and physical processes that occur in the environment. The ability to relate environmental quality to emission sources quantitatively is of central importance in the development of cost-effective methods of pollution control.

The complex relationship between chemical releases and their environmental consequences and the possibility of unexpected side effects are now recognized and feared by the public. Heightened public concern is forcing engineers to change long-established patterns of engineering design.

RETHINKING DESIGN FOR ENVIRONMENT

The unexpected environmental consequences of pollutant releases, the cost of pollution control (currently estimated to be $115 billion annually, or 2.1 percent of GNP, for the United States alone), and continuing population growth and technological development are causing engineers to rethink the basic concepts of engineering design, as they relate to environmentally compatible technologies. There is a growing trend toward waste reduction, or pollution prevention, stressing total system design (as opposed to add-on devices or end-of-pipe treatment) to reduce or avoid waste formation in industrial systems.

In the past, the design of clean technologies has generally meant the use of end-of-pipe treatment or separation devices through which effluent gases or liquids pass on their way to the environment. These devices—electrostatic precipitators, scrubbers, filters, biochemical treatment systems, etc.—are designed to meet government emission standards for particular chemical compounds. They are frequently incorporated after the main features of the technology are designed.

The case for the primacy of pollution prevention over end-of-pipe treatment rests on several factors: avoiding formation of a waste eliminates the need for treatment and disposal, both of which carry environmental risk. Control technologies may fail or fluctuate in efficiency. Treated effluent streams carry nonregulated residual substances that may turn out to be harmful as a result of unanticipated environmental processes. Secured waste disposal sites eventually discharge to the environment.

Although pollution prevention is an attractive concept, the elimination of manufacturing wastes ("zero discharge") is beyond the capability of modern technology. It is a limiting goal that has, however, been adopted by several major chemical companies. The issue is really how to approach this limiting goal in an expeditious and cost-effective manner. Moreover, we must move beyond the "low-hanging fruit," the relatively simple clean-up and housekeeping improvements that have provided much of the past success in waste reduction.

Proponents of pollution prevention (e.g., National Research Council, 1985) have made good use of case studies to illustrate the concept and its application to engineering practice. Various industry sectors have built up a broad base of such studies for practical use. Now, however, we need to find more general scientific and engineering principles on which to base a systems approach to pollution prevention. Generic approaches are of great value in engineering design and badly needed in the organization of research in support of pollution prevention. Generic approaches are essential in engineering education. Engineers are normally taught broadly applicable scientific disciplines, such as thermodynamics and fluid mechanics, which they apply to the solution of design problems.

Basic engineering principles for end-of-pipe treatment and disposal technologies are well estabished. For example, chemical destruction methods are commonly based on combustion and biochemical (microbial) processes. Separation technologies employ filtration, electrical precipitation, scrubbing, and other recognized physical and physicochemical processes to collect or concentrate wastes before destruction or disposal. Although many technical problems remain, a generally accepted framework exists for guiding research and development to improve the performance of waste destruction and separation technologies. Standardized procedures exist for the design of these systems.

One of the most useful basic design paradigms for pollution prevention comes out of chemical process engineering (National Research Council, 1988). Using material balances, thermodynamics, chemical kinetics, and transport phenomena, we can make very accurate estimates of the rates of conversion of reactants or raw materials to desired products in chemical processes and manufacturing plants (refineries, for example). Yields of desirable products may range from 40 or 50 percent to well over 90 percent depending on the particular process. Skilled design engineers can usually predict these yields fairly closely. Yields of undesirable by-products (pollutants) are much smaller. An example is given in Figure 2 in the form of a rank/order diagram for both desirable products and pollutants from a 50,000 barrel per day petroleum refinery. The figure is based on a composite of data gathered from a variety of sources. Figure 2 shows that undesirable by-products are extremely small relative to desired products and raw material (reac-

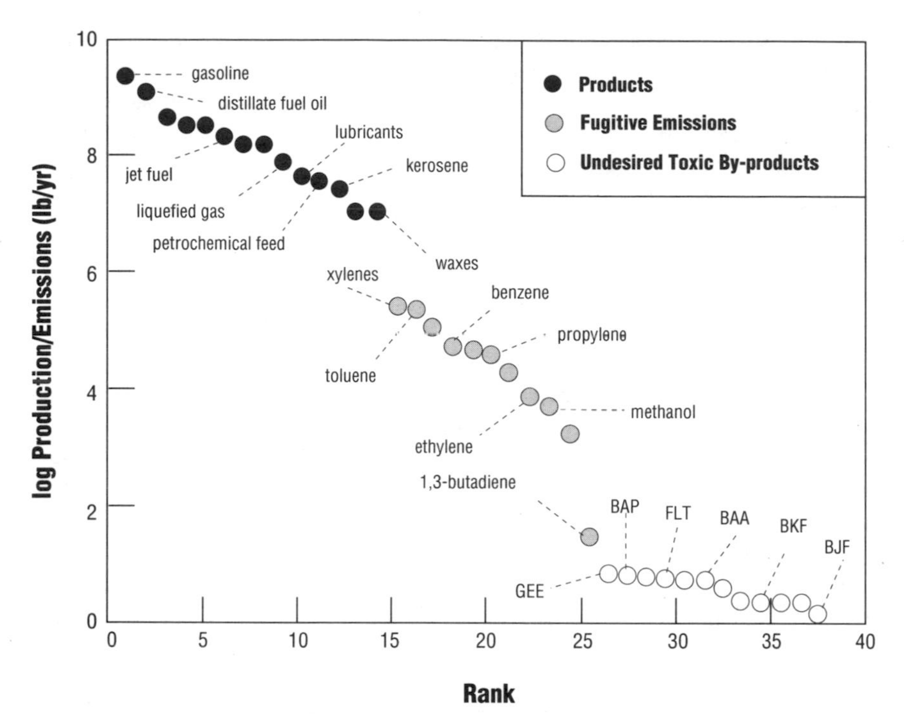

FIGURE 2 Rank/order diagram for outputs from a 50,000 barrel/day petroleum refinery. These estimates are based on composite data from several sources and should be considered illustrative. Normal chemical engineering design procedures based on material and energy balances and chemical kinetics work well in predicting the outputs of desired products. Fugitive emissions of products are difficult to estimate. Releases of pollutant by-products such as polycyclic aromatic hydrocarbons (PAHs) are usually the most difficult to measure or predict. These PAH estimates were based on a material balance using atmospheric data on PAHs and lead with information on PAH and lead emissions in automobile exhaust gases.

tant) inputs to the system. Often, we do not have a good understanding of the mechanisms by which these by-products form, and, sometimes, how they are released to the environment. As a result, the classical engineering approaches, based on material balances, are seriously limited in tracking the sources of undesired by-products. This is true at all scales of industrial organization shown in Figure 1.

All of this means that we will have to sharpen our chemical engineering design skills by several orders of magnitude to go from the raw materials (reactants) to products along pathways of minimum production of undesirable (toxic) by-products, dispersed molecularly or as fine particles. Chemical reaction engineering and recycle theory will take their place along with separation technologies as key engineering sciences for pollution control. This will require major advances in such fields as applied chemical kinetics and in the dynamics of fine particle formation in gases and liquids. There is no shortcut around the point so well articulated by C. P. Snow in 1971: "The only weapon we have to oppose the bad effects of technology is technology itself."

Advancing Snow's notion of "the rational use of technology" for pollution prevention will inevitably involve regulatory action (see Housman and Weinberg et al., in this volume). Finally, we need to give highest priority to engineering education, both in systematizing knowledge and in educating the next generation of engineers in pollution prevention basics.

TRENDS IN ENGINEERING EDUCATION

Industry is actively pursuing pollution prevention using case studies and pragmatic, ad hoc approaches. Academic engineering's role should be to systematize this approach to facilitate its application and to introduce it into engineering education. Design for pollution prevention should be a key feature of the education of the next generation of engineers. The concept should be introduced early in the undergraduate engineering curriculum, based where possible on the principles of engineering science. As appropriate, case studies can be used on a limited basis.

Courses related to the selection and design of manufacturing processes and products should incorporate environmental compatibility from the start, along with scientific and economic factors. As a first step, this can be accomplished by appropriate homework and design problems in existing core courses (see, for example, Allen et al., 1992). This is best done within the framework of the existing engineering disciplines. There is no need for a new branch of engineering to deal with the design of environmentally compatible technologies. Indeed, it would be self-defeating for the development of pollution prevention to fail to inculcate all engineering students with these concepts.

For the chemical process engineer, the challenge is to select raw materials and physicochemical pathways that minimize the formation of toxic by-products. The chemical process engineer with sharply enhanced skills in pollution prevention can serve in all branches of industry—not only the chemical and petroleum and related industries—as the designer of clean chemical technologies. Product engineering, particularly as it relates to the "take-back" mode being pioneered in Europe, will require a somewhat different set of skills.

Although engineers focus on technological systems, they should have a good understanding of the interaction between the technological and environmental systems at various scales of industrial organization (Figure 1). Engineers must be able to meet specialists in public health and ecology at least halfway to put together as complete a picture as possible of the environmental effects of technological systems. For this they need a good understanding of the ecological or public health basis of the regulatory standards that apply to the systems they are designing.

While engineering students are attracted by the idea of developing environmentally compatible technologies, they are also attentive to the signals sent by industry. Industry has a stake in encouraging students to acquire the skills in engineering design needed to protect the environment. Industry should make clear its commitment to the systematic incorporation of pollution prevention in engineering education through participation in the professional societies, in procedures for the accreditation of university curricula and in serving on academic engineering advisory committees.

The design of clean technologies offers the United States an opportunity to obtain a competitive advantage on world markets. The U.S. market is so large that the demands placed on processes and products to satisfy our own environmental regulations have a major impact on technology worldwide. It is in our national interest to remain the leaders in identifying existing or potential undesirable environmental effects and to develop regulations based on special U.S. geographic and demographic requirements and the political factors peculiar to our federal system and its relationship to industry. Surely U.S. industry is in a better position to respond to such regulations than foreign industry. To this end, the nation will need an engineering profession highly skilled in advanced methods for designing clean technologies, through education and basic research.

SUMMARY

Changing perspectives on the interface between technology and environment are causing engineers to rethink how they design industrial systems for environmental compatibility. This is in response to a shifting national consensus that has deep historical roots. There is a growing emphasis on system design for pollution prevention rather than end-of-pipe treatment and disposal. To this end, we seek fundamental engineering principles on which to base design, research, and education for pollution prevention. Highest priority should be given to educating the next generation of engineers in pollution prevention basics. Taking the lead in pollution prevention offers the United States an opportunity to develop an important competitive advantage given the size of its markets and its leadership in developing and setting environmental standards.

ACKNOWLEDGMENTS

I wish to thank Dr. Alan Kao, a Ralph M. Parsons Fellow in Pollution Prevention at UCLA, who helped with the preparation of the figures.

REFERENCES

Allen, D. T., N. Bakshani, and K. S. Rosselot. 1992. Pollution Prevention: Homework and Design Problems for Engineering Curricula. New York: American Institute of Chemical Engineers.

Ayres, R. U. 1989. Industrial metabolism. Pp. 23-49 in Technology and Environment, J. H. Ausubel and H. E. Sladovich, eds. Washington, D.C.: National Academy Press.

Franklin, B. (1744) in L. W. Labaree, ed. 1960. The Papers of Benjamin Franklin, v. 2, January 1, 1735 through December 31, 1744. New Haven, Conn.: Yale University Press.

National Research Council. 1985. Reducing Hazardous Waste Generation: An Evaluation and a Call for Action. Board on Environmental Studies and Toxicology. Washington, D.C.: National Academy Press.

National Research Council. 1988. Frontiers in Chemical Engineering: Research Needs and Opportunities. Board on Chemical Sciences and Technology. Washington, D.C.: National Academy Press.

Snow, C. P. 1971. Public Affairs. New York: Charles Scribner's Sons.

Thoreau, H. D. (1843) as quoted in Philip Van Doren Stern. 1970. The Annotated Walden. New York: Clarkson N. Potter, Inc.

The Greening of Industrial Ecosystems. 1994.
Pp. 228–240. Washington, DC:
National Academy Press.

Industrial Ecology and Design for Environment: The Role of Universities

JOHN R. EHRENFELD

In the wake of the United Nations Conference on Environment and Development (UNCED) in Rio de Janeiro in June 1992, most press notices painted a gloomy picture of the outcome, especially the lack of implementable agreements. Unquestionably, however, UNCED represents a watershed in the continuing move to establish *sustainability* as the basic underlying theme for environmental policy and management strategies. Sustainable development has been defined in several ways, but all are similar to the definition appearing in the report *Our Common Future* (World Commission on Enviroment and Development, 1987), which set in motion the process leading to the Rio conference. That report defines the notion simply as "development that meets the needs of the present generation without compromising the ability of future generations to meet their own needs."

The very simplicity of this watchword belies the complexity and difficulty in achieving such a state, particularly given present societal activities that can hardly be called sustainable, whether they be in the industrialized North or the developing South. Arguments about the ability of the planet to support a growing human population have been with us since Malthus and continue to be debated at length.[1] The economic determinists argue that the system is self-correcting. As soon as natural resources appear to be insufficient to satisfy current needs, new technologies will emerge, driven by economic interests and the inventive spirit of man. This approach, which assumes that the "right" set of technologies will simply show up at each historical juncture, is problematic because the environment that sustains life has certainly become impoverished along the way. Many issues are raised by this failure, but one conclusion seems clear—we, as a global society or a set of sovereign states, must invent a better way of designing and making choices among technologies. I limit my comments here to technological change, not to the need for a concomitant fundamental change in human values and our stance toward nature and the environment.

In the past several decades, societal concerns about environmental protection have led to the creation and implementation of environmental policies, largely in the form of end-of-pipe, technical regulations designed to limit waste flows into the environment. The cost of implementation, and of large penalties imposed on firms that used inappropriate disposal practices in the past, has led to technical innovations and practices that are more effective, but not in any systematic, sustainable way. Relatively systematic, holistic frameworks for analysis and choice, such as product life cycle analysis, are finding their way into practical applications, but as research at the Massachusetts Institute of Technology indicates, quite slowly and with uncertain results (Sullivan, 1992). Other authors in this volume address a new, broader idea called industrial ecology or industrial metabolism, which includes such concepts as dematerialization, design for environment, clean technology, environmentally clean or environmentally conscious manufacturing, and life cycle analysis. All have a technological context; the need is to develop a framework that will lead to sustainable choices of technology, regardless of what it is called.

Today all such concepts are embryonic. For the sake of simplicity, I will use the term *industrial ecology.* It is not yet clear what we mean by industrial ecology or the related concepts, nor how we can apply them to the design and implementation of sustainable development. One view expressed by several authors is that of a largely analytic framework that serves mostly to identify and enumerate the myriad flows of materials and technological artifacts within a web of producers and consumers (Ayres, 1989; Frosch and Gallopoulos, 1989). This aspect of industrial ecology has been called industrial metabolism (Ayres, 1989), but is only one possible way of thinking about this new framework. The idea of an industrial ecology can be expanded to include the institutions that are involved in the technological evolution. If the framework does not incorporate such an institutional aspect, it is not likely to be useful in serving as a practical guide toward sustainable development. Analysis is a necessary part of the overall calculus but not a sufficient framework to guide real decisions and implementation. Elements bearing on economic, legal, political, managerial, and other social processes must also be included. Allenby (1992) refers to this as a metasystem (see Figure 1; see also Tibbs, 1992.)

Other authors in this volume expand the context for developing an industrial ecology framework. The key points include the following:

- The problems appear at several scales that are often coupled: global warming, ozone depletion, urban-smog/lead poisoning, and indoor air pollution.
- Potential solutions appear at different scales and are often closely coupled:

macropolicy choices such as priorities and strategies, sectoral/infrastructural (interfirm) choices, and product/process (intrafirm) choices.

- The system is knowledge-intensive.
- Many values are conflicting.
- Existing rules and structures are conflicting and uncoordinated.
- Technological change is central.
- New forms of learning are necessary.

The development of such a framework is in itself a "daunting task." (Allenby, 1992, p. 63). This is particularly true if the framework is to be created out of whole cloth and by the players most involved in the very decisions that must be made. The key institutional players have become blind to the systems character of the industrial ecological world. This blindness to the complex, interlinked relationships among social units and the natural world, as well as our Western penchant for acting only after creating singular problems to be solved, has produced an atmosphere of blame and adversariness as well as policies that are shortsighted and limited in effectiveness.[2] Two factors, however, support an optimistic outlook toward the development and application of industrial ecology as a means to deal with many of the contextual complexities and challenges.

First, many of the pieces that could be woven together to form an industrial ecology framework are already available in some stage of development. They need, however, to be understood and aggregated. Second, the university can and must play a central role in developing the concept of industrial ecology and institutionalizing its practice. Only the university offers the possibility of a competent institution that has not become blinded or coopted by the current policy and management decision-making system. This is not to say that the university is an ideal nursery, as its own institutional perspective is clouded by strong disciplinary barriers and jealousies and by its own political dynamics. Nevertheless, a practical industrial ecological way of thinking about policy, design, and the world is most likely to arise out of an academic setting, but only if universities are willing to enter into the broad discourse among all the players and to reconstruct the disciplines in a way that mimics the seamless web of the very world that we are attempting to understand. "Seamless web" has become a key metaphor in analyzing the development of large-scale sociotechnological systems, which systems are similar to the larger industrial ecological systems.[3]

The next sections of this paper present elements of an industrial ecology research program and a discussion of the roles that a university might play in such a program.

A PUTATIVE INDUSTRIAL ECOLOGY FRAMEWORK

Figure 1 shows the systems-like character of industrial ecology. Figure 2 presents another view depicting how the concept of industrial ecology might be put into effect to guide technological and other policy and management decisions. A similar concept is being developed at the Massachusetts Institute of Technology (MIT) as part of a current research program to examine chlorine-containing chemicals that form a large family of some 15,000 products in commerce today. The objective of the project is to develop a process to lump the chemicals into a small number of families that can be examined for potential policy options that would phase out, restrict, mitigate, or otherwise change present patterns of production and use. In particular, the project is intended to identify the most promising opportunities for the innovation and application of clean technologies. There is no claim that this is the "right" way to capture the notion of industrial ecology in a practical framework, but it can serve as a model with which to develop a research agenda.

Making sustainable technological choices must combine the best knowledge we can get about the technical and environmental consequences of the option being considered, and our understanding of the workings of social institutions and value systems. Both domains contribute to the processes that create the future. By starting with a more or less traditional model in which the analysis comes first,

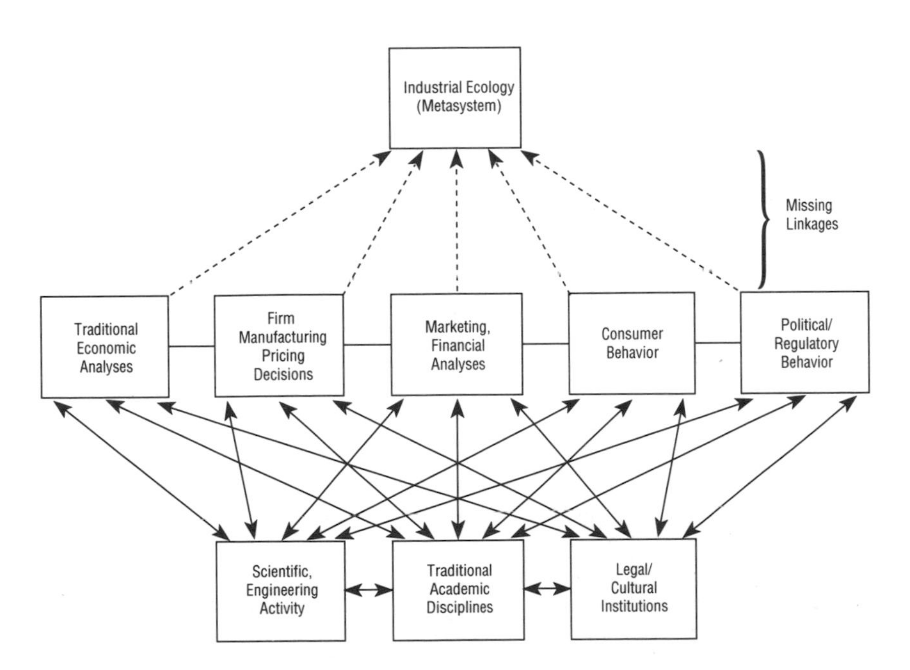

FIGURE 1 Industrial ecology as a metasystem. SOURCE: Allenby (1992).

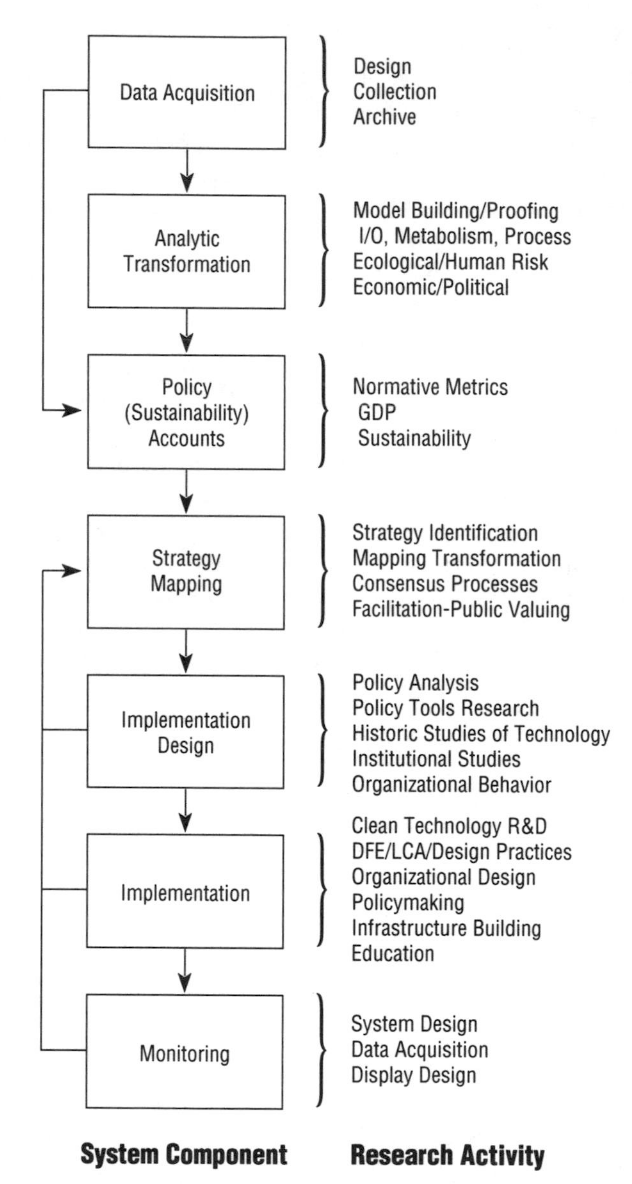

FIGURE 2 Industrial ecology conceptual model.

it will be necessary to map the results into a framework that can guide social choices among policy and managerial options. Given the extreme complexity of such choices and the uncertainty that is always present under these circumstances, a framework that can quickly limit the number of strategic options has much practical value.[4] If such a framework can be developed, then the industrial ecological system can be divided into a set of data-gathering and analytical steps. These are followed, first, by a mapping that transforms the complex information set into a relatively parsimonious set of strategic or policy options and second, by the development of particular implementing mechanisms. Implementation, even if it proceeds according to the agreed upon strategies and mechanisms, is certainly going to introduce unanticipated consequences. Given the seriousness and potential or practical irreversibility of some of the unwanted consequences, some sort of monitoring and feedback should be included as part of the overall industrial ecology framework. Such elements have become part of modern notions of productivity and competitiveness. For example, monitoring, continuous adaptation, and change are fundamental elements of the emerging concepts of total quality management and its environmental counterpart, total quality environmental management.

The following paragraphs further define each of these steps, or stages, and indicate some of the challenges that can shape future academic and other research. I have been guided in this discussion by the work we have been doing to build an industrial ecology model of chlorine.

Data Acquisition—The concept of industrial ecology requires data on the production and consumption of materials and how these are interconnected so that a food-web-like understanding can be established. Economic data at each node in the web are important to evaluate the social impact of change. Jobs are likely to be lost or shifted along with other economic flows. Consumer satisfaction may be forgone or shifted to other technologies. Altenative production and user practices must be specified. It is important to have data that describe the industrial and other institutional organizations involved. Existing policies that constrain choice, such as applicable regulations, and that influence flows in the web, such as subsidies, must be carefully catalogued.

With only a partial discussion of the data that will be needed to construct an industrial ecology, it should be evident that this is a daunting task. Our experience with the chlorine project indicates that publicly available data are limited. Cooperation among private and public institutions will be essential to produce an adequate data base. One important research project might be to develop linkages among separate data bases using the rapidly emerging open architecture and networking capabilities of computer systems.

Analytic Transformation—The data must be manipulated and combined into useful measures for guiding choice. Input/output models such as suggested by Ayres (1992) or Duchin (1992) could be used to collapse data into a smaller set of policy-analytic metrics. Forms of life cycle analyses that can identify and link environmental consequences of technological choice are needed. Alternative economic accounting methods that include environmental cost surrogates reflecting long-term effects and changes in inventories and quality are now being developed. These are but a few of the kinds of analytic systems employing new forms of environmental calculus that will be needed to put the concept of industrial ecology into a practical context.

Policy (Sustainability) Accounts—In a manner perhaps comparable to the use of macroeconomic accounts such as gross national product or gross domestic product in shaping macropolicy choices in most countries today, a new set of sustainability accounts might be developed. The development of such accounts is essential to clarify the idea of sustainability and to develop performance measures by which we can monitor our progress toward achieving a sustainable world. No such standards, agreed upon by more than a handful of people, exist today. Such accounts, which would comprise a much richer set of metrics than the standard economic accounts, might include the following elements:

- Ecological summary, including characteristics of the production/consumer web such as total flows among sectors; materials "value in use" assessments that can assist trade-off comparisons; and others.
- Environmental performance indicators, including estimates of impacts on the environment with measures of the certainty of the assessments. In all of the accounts, estimates of uncertainty are critical, particularly if the so-called precautionary principle is to become a standard part of the industrial ecological policy setting algorithm.
- Technology balance sheet, indicating the maturity of existing products and processes and some measure of potential for innovation.
- Infrastructure balance sheet, indicating the degree to which social and technical infrastructure is adequate to support, or can adapt to, changes.
- Economic indicators, including conventional as well as environmentally enhanced measures.
- Political balance sheet, including the various interests affected by the set of issues under consideration with some assessment of the distribution of power.

Strategy Mapping—This element would consist of some way of using the accounts to locate the set of issues on a policy or strategic grid. For example, Charles Perrow, in his study of complex systems (1984), developed a simple mapping process to point to social choices. He set up a matrix with catastrophic potential and cost of alternative technologies as the two axes and located technologies such as nuclear power and biotechnology on the grid. In an analogy to the precautionary principle, he selected out those technologies he considered likely to fail at some time with consequent unacceptable results. No matter what form of mapping is developed, the process will be strongly influenced by political interests. The design of this part of the overall scheme must incorporate mechanisms to resolve the differences among the many players.

Implementation Design—For each major strategic option, such as phase out, restriction, mitigation, or substitution of clean technologies, a set of implementing mechanisms must be selected and designed. Mechanisms would include conventional policy instruments such as regulation, taxes, and market mechanisms. Given the strong emphasis on technology, new mechanisms that enhance technological change in the directions indicated by the process, without actually defining the technology, will be needed. Current experience with technology-forcing policies and managerial principles indicates that there is much room for innovation and improvement. Again, the choice among alternatives will be strongly influenced by political and other interests, and this step in the overall decision process must incorporate means for bringing in these parties and producing agreements.

Moreover, sustainability calls fundamentally for a new approach to the design of technology and for new public and private institutions. Establishing such institutions and practices must be considered in designing implementation systems.

Implementation—Implementation is largely the responsibility of public- and private-sector institutions other than the university. One activity, research and development of clean technological substitutes for current processes and products, is clearly an area in which academia can play a key and continuing role. Education is another activity that fits into the implementation phase as well: all students must be educated to be environmentally aware, and professionals must be educated to carry out the overall analysis and policymaking process. Shifting curricula toward a new paradigm will not be an easy transition for schools deeply steeped in old ways of thinking, but it is a necessary shift all the same.

Monitoring—Given the complexity of the industrial ecological system and the incremental impact of many potential polices, it is important to monitor progress toward sustainability, learning more about the notion in the process. Continuous learning will be necessary. Learning in the systems sense will require some sort of closely coupled feedback between the acting and designing parties and the world as it is affected by the actions taken. Environmental monitoring has been grossly ignored historically. New technological systems and institutions to develop and handle the data on a global basis must be developed. Both will require considerable research.

SELECTED RESEARCH EXAMPLES

The foregoing discussion points to general areas of academic research and includes a number of suggested topics for universities (see the right-hand column in Figure 2). The set of categories does not easily identify other research areas that may be cross-cutting or subjects of particular concern. The following short list is included to provide additional examples: it is not exhaustive, but only representative of the large family of research that must be done to develop such an industrial ecology framework. The point being made is that there are unanswered questions that fit well into the academic research structure.

Data Acquisition—Amassing the quantities of data required to support a large industrial ecology system is a huge task. Many key data exist, but in many forms and places. Other important data do not exist and must be developed. A key research project might be to develop data networks that can access and combine such disparate sources of information. New forms of data bases must be designed and built to accommodate the demanding analytic needs of an industrial ecology framework. The availability of large relational data bases should avoid the task of creating data systems de novo.

Analytic Transformation—The work begun by several researchers on industrial metabolism models requires considerable further development. Ayres and Duchin (in this volume) provide examples of different approaches that can be developed to indicate and quantify relationships between material flows across industrial sectors. Research on life cycle analysis also provides a systematic framework for analyzing product and process implications.

Policy Accounts—Little progress will be made over time unless sustainable performance measures are developed. The current economic system of accounts now widely used to guide policy decisions does not capture factors deemed important for long-term decision making. Accounting and management information systems include expansions of standard macroaccounting systems used at national and international policy levels and microsystems used within the firm for management decisions (see Todd, in this volume). In addition, life cycle frameworks that can provide information on the environmental future of contemplated actions are critically important.

Strategy Mapping—New and improved processes for negotiating and agreeing on values and criteria are needed. The earlier discussion should have made clear the political context of industrial ecology. No progress is likely to be made without some consensus among the many interests involved that the policies and strategies are necessary and ultimately serve their interests. Better means and institutional frameworks for introducing science into policy dialogues and for dealing with uncertainly are high priority areas. As the ecological web grows larger and the time frame becomes longer, more and more uncertainty will show itself to the players. Finding ways to avoid stagnation and vacillation under these conditions is critical.

Implementation Design—Institutional studies of firm behavior are very important. Many alternative clean technologies exist today but are not being used by firms in spite of potential cost savings and product quality gains. Research on the reasons for such apparently irrational behavior is needed, together with other fundamental studies of innovation and design practices. Although this field is large and well established, current understanding is inadequate to suggest changes that should produce a more sustainable result. Organizational learning is another key area, as new demands will require large shifts in firm culture and competence. A recent column in *Environmental Science and Technology* (Alm, 1992) notes that social science, particularly that leading to an "understanding of how industrial firms make decisions," has been notably absent from the research program of the U.S. Environmental Protection Agency. The field of historical and sociological studies of technology offers great opportunity for understanding the broad systems dynamics of technological change and its relationship to social institutions. Recent work on large-scale systems is of special interest.

Implementation—New "clean technologies"—processes and products to replace the particularly "dirty" current technologies—are clearly needed. Chemical, fiber, and metals production technologies, for example, have evolved under a set of constraints that have created large quantities of waste. New (and, perhaps, some old) ideas are needed. Institutional forms that are more cooperative and integrate players across sectors and national boundaries need to be designed and studied.

Monitoring—Given the complex and often highly uncertain nature of the problems that will fall under an industrial ecology rubric, new forms of monitoring progress are critical. The present approach to environmental decision making often assumes relatively certain answers to the problems posed and proceeds with little provision for monitoring the actual consequences. To the extent that it occurs, monitoring frequently targets environmental proxies, such as compliance with regulations, but fails to follow the environment itself. New technologies and monitoring strategies are needed to address the daunting questions raised by many current and prospective concerns.

These are but a few elements of an academic research agenda. The research needed is substantial, calling on new alliances among research disciplines and perhaps creating new fields.

CLIMBING DOWN FROM THE TOWER

The policy and management context of industrial ecology, as presented in this paper, suggests that the university must adopt a more active stance than its traditional detached, scholarly perspective. Even the preparation of a carefully developed set of research papers presenting new analytic frameworks and implementation policies and strategies is not likely to have a significant immediate effect. Traditional academic audiences are narrow, self-centered, and organized by discipline, whereas the study of industrial ecology should be broad and involved in practice. To be effective in moving toward sustainability, the research agenda must be linked to activities designed specifically to intervene in current practices and to change institutional forms. Industrial ecology is, at least in part, a new way of thinking; it is not merely the expansion of an existing set of theories about the world. Thus, if change is to occur, it must include cultural and institutional transformations. The university can contribute to such shifts by exercising its historical role in education and by encouraging and becoming more directly involved in public policy dialogues, a less traditional and often perilous role.

EDUCATION

Education is one way that the new ideas and ways of thinking can be diffused into the world. It is important that universities introduce the industrial ecology notion as a fundamental framework in teaching subjects of all kinds. This notion should be part of the general thrust toward improving environmental literacy. Much of the focus on literacy has been on learning how the environment works and how important it is to human activity. I am suggesting that it is also important to include a large dose of industrial ecological context so that students begin to understand how intimately virtually everything we *do* is tied to the natural world.

The need to educate professionals in fields, such as business administration, where environment has been virtually ignored is now being recognized. Although such professionals will have to spend much of their time coping with existing regulatory demands, it is important to instill the broader context of industrial ecology. Otherwise, we are likely to continue to seek solutions to problems rather than to avoid them in the first place. The technological consequences of societal activities should become a more explicit part of the education of professionals heading for planning, policy, managerial, and design careers. Education of professionals already in practice is also important, as this group is most directly involved in the decisions that influence technological choice.

INTERVENING IN THE PUBLIC DIALOGUE

Industrial ecology is a manifestation of the recognition that human activities and the natural world are inextricably linked. This linkage joins the so-called objective scientific set of disciplines with the social, practical world. In that practical world, individual and societal values enter into the decision and design processes that continuously shape the future. It is impossible to conceive of a purely objective, detached way of thinking about sustainability, far from the reality of everyday activities. The most important decisions are those made by the players. The university needs to join that circle to be able to bring its knowledge and systems purview to full fruition and effectiveness. The road to sustainability, as the preparations for UNCED 1992 demonstrated, is obstructed by disagreements among the major interests. The university, building on its objective research base, can play a key role in creating an action-producing discourse among the players. Much theory points to the essential need for trust and a sense of legitimacy in consensus-building processes. The university, as an institution, appears to have largely maintained its legitimacy as an objective and competent institution while public and private bodies have been battered of late.

By convening and facilitating dialogue on key issues, universities can forward the action, far beyond the normal path of publishing papers. Individual faculty members have long taken stands on issues. I speak of a different process in which the whole institution acts as a neutral party. Its role is not to take a stand on the issues out of some interest beyond its legitimate bounds, but to point out the state of knowledge of the complicated world and to compare the many options always available to decision makers. We, at MIT, have had some success in this kind of endeavor (Ehrenfeld et al., 1989) and are planning to continue our attempts at helping parties now at odds to come to agreement, even if very slowly. This may well offer a model for future emulation.

NOTES

1. One modern notion of limits grew out of a study using systems dynamics to model global processes and was reported by D.H. Meadows and coauthors (1972) in *The Limits to Growth.* Some 20 years later, the same project team (Meadows et al., 1992) reiterated their thesis in *Beyond the Limits.* The opposite viewpoint has been most vociferously argued by Simon and Kahn (1984) in *The Resourceful Earth: A Response to Global 2000.*
2. Peter Senge (1990) in his recent book on organizational learning, *The Fifth Discipline*, has observed that failures to perceive the systems context lead to a problem-solving mode that first must find someone or some organization to blame. This interesting observation may help explain the extreme degree of adversariness that surrounds environmental decision making in the United States. The well-developed technical, rational framework and clear disciplinary bounds that characterize administrative and managerial processes in most of the highly industrialized nations has contributed to the blinders that constrain public and private strategic activities.
3. See Ehrenfeld (1990). This paper refers to the sociology of technology, and in particular to a quote from Thomas P. Hughes (1988). Hughes, notes that: Callon asks why we categorize, or compartmentalize, the elements in a system or network "when these elements are permanently interacting, being associated with, and being tested by the actors who innovate?" Faced by the rigid-categories problem—science, technology, economics, politics, etc.—Callon resorts to neologising and uses higher abstractions (actors) that subsume science, technology and other categories. Actors are the heterogeneous entities that constitute a network. Disciplines do not bound actors. The historian or sociologist using the expression need not introduce connotative terms such as the political, social, or economic.

   As a case study, Callon employs the post-World War II effort of the French state to promote an electric vehicle. His actors include electrons, catalysts, accumulators, users, researchers, manufacturers, and ministerial departments defining and enforcing regulations affecting technology. These and many other actors interact through networks to create a coherent actor world. Callon does not, therefore, distinguish between the animate and inanimate, the individuals and the organizations. He sees no outside (social) - inside (technology) dichotomy.

   The passage quoted above refers to two papers (Callon, 1980; Callon and Law, 1982).
4. See Lindblom (1988, pp. 237–259, for example. Achieving impossible feats of synopsis is a bootless, unproductive ideal. Aspiring to improving policy analysis through the use of strategies is a *directing* or *guiding* aspiration. It points to something to be done, something to be studied and learned, and something that can be successfully approximated.

REFERENCES

Allenby, B. R. 1992. Achieving sustainable development through industrial ecology. International Environmental Affairs 4(1):62.

Alm, A. 1992. Science, social science, and the new paradigm. Environmental Science and Technology 26(6):1123.

Ayres, R. U. 1989. Industrial metabolism. Pp. 23-49 in Technology and Enviroment, J. H. Ausubel and H. E. Sladovich, eds. Washington, D.C.: National Academy Press.

Ayres, R. U. 1992. Toxic heavy metals: Materials cycle optimization. Proceedings of the National Academy of Sciences 89:812-820.

Callon, M. 1980. The state and technical innovation: A case study of the electric vehicle in France. Research Policy 9:358-376.

Callon, M., and J. Law. 1982. On interests and their transformation: Enrolment and counter-enrolment. Social Studies in Science 12:615-625.

Duchin, F. 1992. Industrial input-output analysis: Implications for industrial ecology. Proceedings of the National Academy of Sciences 89:851-855.

Ehrenfeld, J. R. 1990. Technology and the environment: A map or a mobius strip? Paper prepared for Toward 2000: Environment, Technology, and the New Century, a World Resources Institute symposium, Annapolis, Md., June 13-15, 1990.

Ehrenfeld, J. R., E. P. Craig, and J. Nash. 1989. Waste incineration: Confronting the sources of disagreement. Enviromental Impact Assessment Review 9:305–315.

Frosch, R. A., and N. E. Gallopoulos. 1989. Strategies for manufacturing. Scientific American 261(3):144-152.

Hughes, T. P. 1988. The seamless web: Technology, science, et cetera, et cetera. In Technology and Social Process, B. Elliott, ed. Edinburgh, Scotland: Edinburgh University Press.

Lindblom, C. E. 1988. Democracy and the Marketplace. New York: Oxford University Press.

Meadows, D. H., D. L. Meadows, J. Randers, and W. W. Behrens III. 1972. The Limits to Growth. New York: Universe Books.

Meadows, D. H., D. L. Meadows, J. Randers, and W. W. Behrens III. 1992. Beyond the Limits. Post Mills, Vt.: Chelsea Green Publishing.

Perrow, C. 1984. Normal Accidents. New York: Basic Books.

Senge, P. 1990. The Fifth Discipline. New York: Doubleday.

Simon, J. L., and H. Kahn, eds. 1984. The Resourceful Earth: A Response to Global 2000. Oxford: Basil Blackwell.

Sullivan, M. 1992. Enviromental Life Cycle Frameworks: Industry Management of Product Innovation and Environmental Impact. M.S. thesis. Technology and Policy Program, Massachusetts Institute of Technology.

Tibbs, H. B. C. 1992. Industrial ecology—An agenda for environment management. Pollution Prevention Review Spring:167-180.

World Commission on Enviroment and Development. 1987. Our Common Future. New York: Oxford University Press.

Biographical Data

DAVID T. ALLEN is chairman of the chemical engineering department at the University of California, Los Angeles. During the 1989/90 academic, year he was a visiting associate professor of chemical engineering at the California Institute of Technology. Since 1987 Allen has led the Waste Reduction Engineering research effort at UCLA. This program has been one of the few university research efforts directed specifically at waste reduction. In addition, he serves on a number of national advisory committees on waste rcduction. He is a member of the Environmental Protection Agency's American Institute for Pollution Prevention; he serves on the Pollution Prevention Education Group of EPA's National Advisory Council for Environmental Policy and Technology, and is actively developing pollution prevention materials for engineering curricula. Allen received his B.S. degree in chemical engineering, with distinction, from Cornell University in 1979. His M.S. and Ph.D. degrees, also in chemical engineering, were awarded by the California Institute of Technology in 1981 and 1983.

BRADEN R. ALLENBY is the research vice president, technology and environment, for AT&T. He graduated cum laude from Yale University in 1972 and received his Juris Doctor from the University of Virginia Law School in 1978 and his master's in economics from the University of Virginia in 1979. He received his master's in environmental sciences from Rutgers University in 1989 and his Ph.D. in environmental sciences from Rutgers in 1992. Allenby is a member of the Aeronautics Board and the Federal Communications Commission, as well as a strategic consultant on economic and technical telecommunications issues. He joined AT&T in 1983 as a telecommunications regulatory attorney and was an environmental attorney and senior environmental attorney for AT&T from 1984 to 1993. During 1992 he was the J. Herbert Hollomon Fellow at the National Academy of Engineering in Washington, D.C. He currently chairs the American Electronics Association Design for Environment Task Force and is vice-chair of the Institute of Electrical and Electronics Engineers Committee on the Environment.

FREDERICK R. ANDERSON, of the Washington, D.C., law firm Cadwalader, Wickersham, and Taft, is former dean of the law school and Ann Loeb Bronfman Professor of Law at American University. Anderson is a nationally recognized authority on environmental law. In addition to writing, teaching, and testifying on a broad range of matters involving science, natural resources, and the environment, Anderson practices in the areas of air, water, and hazardous substance pollution, the environmental aspects of energy production and real estate development, natural resources management, environmental compliance and cleanup, and administrative law. In 1970 Anderson was the founding president of the Environmental Law Institute and is currently chairman of the Institute's Advisory Council. He was the first editor-in-chief of the *Environmental Law Reporter.* For several years he was chairman of the American Bar Association's Standing Committee on Environmental Law. He served on a twelve-member congressional study commission created by the Superfund legislation to examine toxic tort recovery for injury from hazardous substances. Anderson currently chairs a broad-based advisory group that is attempting on behalf of the U.S. Sentencing Commission to draft sentencing guidelines for individuals and organizations convicted of environmental offenses. He also serves as a member of the National Research Council's Board on Environmental Studies and Toxicology. Anderson graduated summa cum laude from the University of North Carolina with a B.A. in the history of science. He holds law degrees from Harvard University and Oxford University, where he was a Marshall Scholar.

ROBERT U. AYRES was professor of Engineering and Public Policy, Carnegie Mellon University from 1979 through May 1992. In September 1992 he became Sandoz Professor of Environment and Management at the European Institute for Business Administration (INSEAD), Fountainbleau, France. His current research and teaching interests focus mainly on "industrial metabolism" and the impacts of technological change and public policy initiatives in response to environmental concerns. Ayres graduated from the University of Chicago (B.A. 1952, B.S. 1954) in mathematics, then did graduate work in physics at Maryland (M.S. 1956) and Kings College, University of London (Ph.D. 1958). This was followed by postdoctoral research in physics at Maryland and Yeshiva University from 1960 to 1962. In 1962 Ayres joined the staff of the Hudson Institute, where he worked on environmental problems until 1967. After a year as a visiting scholar at Resources for the Future, Inc. (1967/68) he cofounded a research and consulting firm in Washington, D.C., to perform studies for the U.S. government. He has published more than 120 journal articles and book chapters, and has authored or coauthored 10 books and edited several others on topics ranging from technological change, manufacturing and productivity to environmental economics.

NASRIN BEHMANESH is a postdoctoral research associate in the Department of Chemical Engineering at the University of California, Los Angeles. Her current research focuses on pollution prevention, waste reduction, life cycle assessment and alternate uses of hazardous wastes. She has coauthored several journal articles in environmental engineering. Behmanesh received her Ph.D. in chemical engineering from the University of California, Los Angeles.

PIERRE CROSSON is an economist at Resources for the Future, where he has worked for over 25 years. For the last decade or so his research has focused on the impacts of agriculture on the natural resource base and environment, emphasizing the consequences for the sustainability of the agricultural system, in the United States and globally. With colleagues at RFF he currently is engaged in developing a program of research on sustainable development. Crosson received his undergraduate degree from the University of Texas and his Ph.D. in economics from Columbia University.

PATRICIA S. DILLON is a senior research associate at the Tufts University Center for Environmental Management, where she specializes in corporate environmental management issues. Her current research examines product stewardship and related developments in the private sector, such as Design for Environment and industrial ecology. Dillon has written widely on business and the environment and on various aspects of environmental management in corporations. Dillon is a member of the advisory board of the Greening of Industry project, an international research network on corporate environmental behavior. She holds an M.S. degree in civil engineering and a B.S. degree in biology from Tufts University.

FAYE DUCHIN is director of the Institute for Economic Analysis at New York University. Over the past several years, she has investigated the potential contribution of technological change to environmentally sound economic development in the international context and for individual developing countries. Duchin is a founding managing editor of *Structural Change and Economic Dynamics*, a journal promoting the integration of theoretical and empirical research in economics. She is a vice president of the international society for ecological economics, where she is responsible for curriculum development. Duchin received her B.A. degree in psychology from Cornell University and her Ph.D. degree in computer science from the University of California, Berkeley.

JOHN R. EHRENFELD is senior research associate in the Massachusetts Institute of Technology Center for Technology, Policy, and Industrial Development and has additional appointments as lecturer in the departments of chemical engineering and urban studies and planning. At MIT since 1986, he directs the MIT Program on Technology, Business, and Environment, an interdisciplinary educational, research, and policy program. In 1977 Ehrenfeld was appointed by President Carter to serve as chairman of the New England River Basins Commission (NERBC). There he was responsible for developing regional policies and strategies for surface and ground water, and coastal resources. Ehrenfeld has served on the Massachusetts Water Resources Commission, the state's primary water policy organization, and on the boards of other public and nonprofit organizations. He holds a B.S. and Sc.D. in chemical engineering from MIT.

GREG EYRING received a Ph.D. in chemistry from the University of California, Berkeley, in 1981. He then did three years of postdoctoral research in chemical physics at Stanford University before joining the Office of Technology Assessment staff as an OTA Fellow in 1984. He was project director of OTA study *Advanced Materials by Design* (June 1988), which discusses the status and policy implications of advanced ceramic and composite materials technologies. He then directed OTA's assessment *High-Temperature Superconductivity in Perspective* (May 1990), which evaluates the potential applications of high-temperature superconductors and includes an extensive survey of superconductivity research activities of U.S. and Japanese firms. A third study, *Green Products by Design: Choices for a Cleaner Environment* (September 1992), examines how better product design can help to address environmental problems. In particular, the study explores trends in designers' use of materials, and how policymakers can shape environmental policies that encourage environmentally sound design.

SHELDON K. FRIEDLANDER is Parsons Professor of Chemical Engineering at the University of California, Los Angeles, currently working on the applications of aerosol science and technology to pollution prevention and to the formation and behavior of ultrafine particles. From 1984 to 1988 he chaired the UCLA Chemical Engineering Department. He has consulted for the Los Angeles Air Pollution Control District and served as chairman of the National Research Council Panel on the Abatement of Particulate Emissions from Stationary Sources and the subcommittee on Photochemical Oxidants and Ozone. He was also chairman of the EPA Clean Air Scientific Advisory Committee (1978–1982) and a member of its Science Advisory Board Executive Committee. He is a member of the National Academy of Engineering. Friedlander received his B.S. degree in chemical engineering from Columbia University and Ph.D. degree from the University of Illinois.

ROBERT A. FROSCH is a senior research fellow at the John F. Kennedy School of Government, Harvard University, and a senior fellow at the National Academy of Engineering. He recently retired from the position of vice-president in charge of General Motors research laboratories. Frosch's career combines varied research and administrative experience in industry and in government service. He has been involved in global environmental research and policy issues at both the national and the international level. From 1951 to 1963 he was employed at Hudson Laboratories of Columbia University, first as a research scientist and then as director from 1956 to 1963. In 1963 he became director for nuclear test detection in the Advanced Research Projects Agency (ARPA) of the Department of Defense, and deputy director of ARPA in 1965. In 1966 he was appointed assistant secretary of the Navy for research and development. He served in this position until January 1973, when he became assistant executive director of the United Nations Environment Program. In 1975 he became associate director for applied oceanography at the Woods Hole Oceanographic Institution, and from 1977 to 1981 he served as administrator of the National Aeronautics and Space Administration. He served as president of the American Association of Engineering Societies from 1981 to 1982. Frosch is a member of the National Academy of Engineering. He received his A.B., M.S., and Ph.D. degrees in theoretical physics from Columbia University.

ROBERT F. HOUSMAN is an attorney with the Center for International Environmental Law (CIEL) and an adjunct professor of law at the American University's Washington College of Law. Housman serves as a consultant to the Organization for Economic Cooperation and Development, the United Nations Environment Program, and the Environmental Protection Agency's National Advisory Council for Environmental Policy and Technology. He is also a member of the Council on Foreign Relations Study Group on Trade and Environment. Before joining CIEL, Housman was an associate at Skadden, Arps, Slate, Meagher & Flom's Washington, D.C. office. He also served as aide to Dr. Iann Twinn, member of the English House of Commons.

DAVID JENSEN is an analyst with the Energy and Materials Program of the congressional Office of Technology Assessment. Jensen joined OTA in the fall of 1991, after completing his graduate work in the Department of Engineering and Policy at Washington University in St. Louis. He received his D.Sc. and M.S. degrees from Washington University and his B.S. in mechanical engineering from the University of Nebraska. Since joining OTA, Jensen served on the project team for *Green Products by Design: Choices for a Cleaner Environment*, a study of how product design and environmental regulations interact to affect the environmental attributes of products.

RICHARD L. KLIMISCH is vice president of engineering affairs at the American Automobile Manufacturers Association. From 1983 to 1993 he was executive director of the General Motors Environmental Activities Staff, which oversaw worldwide GM activities related to vehicle safety, fuel economy, and all aspects of pollution. From 1975 to 1983, he was head of the Environmental Science Department of the GM Research Laboratories. Klimisch joined the GM Research Laboratories in 1967 as GM's first resident expert in catalysis. After receiving his Ph.D. in chemistry at Purdue University in 1964, Klimisch was employed as a research chemist for the DuPont Co. at the Experimental Station in Wilmington, Delaware. He received his B.S. in chemistry from Loras College in Dubuque, Iowa in 1960. Klimisch is the author of two books and more than 30 scientific papers and holds two patents in chemical kinetics, catalysis, emission control, atmospheric chemistry, and alternative fuels.

HENRY R. LINDEN is the Max McGraw Professor of Energy and Power Engineering and Management at Illinois Institute of Technology (IIT) and director of IIT's interdisciplinary Energy and Power Center. From July 1989 until June 1990, he served as interim president and chief executive officer of IIT and interim chairman and chief executive officer of its subsidiary, IIT Research Institute. He has been a member of the faculty of IIT since 1954 and served as the Frank W. Gunsaulus Distinguished Professor of Chemical Engineering following his retirement as president of Gas Research Institute (GRI) in April 1987, until becoming interim president of IIT. Linden was instrumental in the organization of GRI, the U.S. gas industry's cooperative research and development arm. He was elected GRI's first president and a member of its board of directors in 1977 and continues to serve GRI as executive advisor. Linden received a bachelor's degree in chemical engineering from Georgia Institute of Technology in 1944, a master's in chemical engineering degree from Polytechnic Institute of Brooklyn in 1947, and a Ph.D. in chemical engineering from IIT in 1952. He is a member of the National Academy of Engineering and a member of the Energy Engineering Board of the National Research Council. During the past six administrations, he has served on many government advisory bodies, most recently on the Energy Research Advisory Board of the Department of Energy. He has written and lectured extensively on U.S. and world energy issues, has authored or coauthored 200 publications, and holds 27 patents.

EDWARD T. MOREHOUSE, JR., a major with the United States Air Force, is military assistant to the Deputy Under Secretary of Defense for Environmental Security. He was previously chief of the Pollution Prevention Division with the Air Force's Environmental Quality Directorate, responsible for developing and implementing a Pollution Prevention Program throughout the Air Force. His earlier assignments include base engineering at Offutt Air Force, Nebraska; RAF Fairford, UK; and Galena Airport, Alaska. He was also the Ground Launched Cruise Missile Program Manager at the London Regional Civil Engineering Office, served in an Education with Industry assignment with the 3M Corporation in St. Paul, Minnesota, and worked at the Air Force Civil Engineering Support Agency's research laboratory at Tyndall AFB, Florida, on alternatives to ozone-depleting substances. Since 1987 Morehouse has been active in the United Nations Montreal Protocol process to eliminate global production of ozone layer depleting chemicals. He cochairs the United Nations Halon Technical Options Committee and is a member of the United Nations Technology and Economics Assessment Panel, which prepares assessment used in the Montreal Protocol process. He has conducted workshops on alternatives to ozone-depleting substances in a number of developing countries as part of a United Nations effort to broaden international participation in the Montreal Protocol. Morehouse holds a B.S. in electrical engineering from Union College and an M.S. in mechanical engineering from Boston University.

ROBERT C. PFAHL, JR., is director of advanced manufacturing technology at Motorola, where his organization is responsible for developing and introducing advanced manufacturing technology and advanced electronic packaging into production. His organization's areas of activity include reflow soldering, cleaning, advanced electronic packaging, factory control systems, and advanced electronic materials. Pfahl is chairman of the American Electronics Association's CFC Task Force. He chairs the Motorola corporate task force that eliminated the use of CFCs in manufacturing in 1992 and has been an active participant in the Industry Cooperative for Ozone Layer Protection. He represents the electronics industry on the Safe Alternatives Subcommittee of the EPA's Stratospheric Ozone Protection Advisory Council. Pfahl invented, developed, and implemented "vapor phase" soldering, used for reflow soldering of temperature-sensitive assemblies. He led the development of new manufacturing processes including infrared solder fusing, liquid immersion solder fusing, and flat flexible cable termination. Pfahl received his B.M.E., M.S., and Ph.D. degrees from Cornell University, where he majored in heat transfer and fluid mechanics.

JOE RAGUSO is a research contractor with the Energy and Materials Program at the congressional Office of Technology Assessment. Current projects include a background paper investigating the commercial importance of biopolymers and a full study examining energy technology transfer to former communist countries working to convert from central planned to market based economies. His most recently completed study entitled *Green Design: Choices for a Cleaner Environment*, for the House Committees on Science, Space, and Technology and on Energy and Commerce, examines the environmental implications of trends in materials technology and product design. Before coming to OTA in 1991, Raguso worked for IBM on developing novel computer substrate materials. He received an M.S. in technology policy from MIT in 1991, as well as an M.S. in glass engineering and a B.S. in ceramic engineering from Alfred University.

DEANNA J. RICHARDS is senior program officer with the National Academy of Engineering and directs the Academy's Technology and Environment program.

Before joining the Academy in 1990, she was an assistant professor at the University of North Carolina at Charlotte and in 1989, as an American Association for the Advancement of Sciences Environmental Sciences and Engineering Fellow she worked on remedy selection criteria at Superfund sites. She has several articles published on her engineering research work on advanced biological wastewater treatment systems. Richards, a registered engineer, has also done engineering consulting in Malaysia and the United States. Richards received her B.S. degree in civil engineering from the University of Edinburgh, Scotland, and her M.S. and Ph.D. degrees, also in civil engineering, from the University of Pennsylvania.

JANINE C. SEKUTOWSKI is a technical manager in the Environmental and Materials Technology Department, AT&T Bell Laboratories. Her group is responsible for research to reduce the environmental impact of AT&T's products and manufacturing operations. She joined AT&T in 1979 as a member of technical staff at the Western Electric Engineering Research Center. She became a supervisor in 1982 and has worked in various areas of manufacturing technology, such as plastics processing and interconnection technology. Sekutowski has a bachelor's degree in chemistry from Kent State University and M.S. and Ph.D. degrees in inorganic chemistry from the University of Illinois, Urbana-Champaign. Following her formal academic training, she had a postdoctoral appointment at the Max Planck Institut für Kohlenforschung in Mulheim a.d. Ruhr, Germany, and at Texas A&M University, where she was also a chemistry instructor. Sekutowski has been the author or coauthor of more than 20 publications on her academic and industrial work.

WALTER R. STAHEL is a director of the Product-Life Institute in Geneva, and deputy secretary general of the International Association for the Study of Insurance Economics. His independent business consulting interests are in utilization-related technologies such as reuse, repair, reconditioning, and technological upgrading of components, goods, and systems; risk management and insurance; and regional economic development. For several years he worked as a private architect in the United Kingdom and Switzerland and in 1973 joined the Battelle Geneva Research Center as project manager in applied economics for business strategies and feasibility studies. In 1979 he became personal assistant to the chief executive officer of a holding company with worldwide activities in railway maintenance, shipping, and real estate. Stahel has authored books and articles on strategies for the improved use of resources and job creation. In 1978 he was awarded a first prize in a German competition on job creation and in 1982 was a recipient of the U.S. Mitchell-Prize for his paper on "The Product-Life Factor." Stahel is an alumnus of the Swiss Federal Institute of Technology in Zürich, where he received a diploma in architecture and town planning.

REBECCA TODD is a member of the accounting faculty of the Stern School of

New York University. She received her Ph.D. degree in business from the University of North Carolina at Chapel Hill in 1986. Her interests include the information content in financial information and the development of improved analysis tools to enhance the financial decision making of managers, financial analysts, and investors. Todd's current research is directed toward developing accounting and analysis tools to enable managers to better capture and trace environmental costs. She teaches several courses in financial statement analysis for manufacturing firms, financial institutions, and cross-border firms in the MBA program at the Stern School. She is a chartered financial analyst and teaches financial statement analysis in candidate curriculum programs in the United States, Europe, and the Far East.

MICHAEL A. TOMAN is a senior fellow at Resources for the Future, a nonprofit and nonadvocacy research organization in Washington, D.C., that specializes in a variety of issues relating to natural resources and the environment. He also is a professional lecturer at the Johns Hopkins School of Advanced International Studies, where his teaching responsibilities cover the economics of natural resources and environment. Toman received his Ph.D. in economics from the University of Rochester, with earlier degrees in economics and mathematics from Indiana University and Brown University. He is the coauthor of two books published by Resources for the Future and is the author or coauthor of more than 30 scholarly and popular publications covering a number of topics related to energy, public utilities, and the environment.

MATTHEW WEINBERG is an analyst with the Energy and Materials Program at the congressional Office of Technology Assessment. He recently completed a study for the House Committees on Science, Space, and Technology and on Energy and Commerce, examining the environmental implications of trends in materials technology and product design. Before coming to OTA in 1990, Weinberg spent six years in the semiconductor industry, where he received five patents in the area of microelectronic devices. He received B.S. and M.S. degrees in electrical engineering in 1983 from the Massachusetts Institute of Technology. His graduate work focused on nonequilibrium effects in superconducting materials.

Index

A

Accounting practice
control system components and, 196–197
engineering/manufacturing knowledge in, 191, 193, 200
environmental costs in, 15, 191, 234
environmentally sensitive, 199–200
in international agreements, 113
organizational barriers to environmental sensitivity in, 193–196
public access to information, 194
role of, 193
sustainability accounts in, 234
traditional, 197–199
in waste reduction strategies, 191–193
Air Force Pollution Prevention Program
education/training in, 153–154
funding, 152
green weapon systems, 159–163
incentives for compliance, 154–155
information needs, 163
objectives, 151–153, 163–164
origins, 149–151
ozone-depleting chemicals in, 149, 152, 160
procedural obstacles, 159–160
purchasing procedures, 155–159, 160–163
Antitrust law, 5, 103, 131 n.12
Automobile industry
catalytic converter technology, 36
current recycling practice, 4, 165–167
environmental regulation for, 169–170
in functionality economy, 16
life cycle analysis, 182*t*
life cycle analysis and recycling in, 169
mandatory recycling of used autos, 127
plastics recycling, 168

B

Barcelona Convention, 114
Bell Telephone System, 16
Bottle bills, 115

C

Cadmium, 73–74
Carbon dioxide, 8, 37 n.4
economic modeling of future emissions, 66

in fossil fuel consumption, 40–41, 42–43, 55–56
non-energy production sources, 57 n.1
rate of increase, 123
world fossil fuel emissions, 54
Chemical engineering design paradigms, 218–220, 223–225
Chlorofluorocarbons, 33–34, 37 n.4, 114, 142, 210–211
Chromium
chromated coatings, 209–210
in industrial waste stream, 73–74
Citizen's Clearinghouse for Hazardous Wastes, 120 n.1
Clean Air Act, 43, 102, 104–105, 210
Closed-system material flow, 25–27
Coal energy, 36, 55–56
methane emissions, 42
Coalition of Northeastern Governors, 206
Commerce, Department of, 129
Concurrent engineering, 11, 12
Consumer protection laws, 5
Consumerism, environmental, 140, 165
in government purchasing, 163
in product design process, 173
Convention for the Prevention of Marine Pollution from Land-Based Sources, 114
Copper, recoverability, 78
Cross-functional teams, 12, 204

D

Defense, Department of, 149, 150–151, 160
Deforestation, 57 n.1
Design for Environment, 14–15
AT&T telephone, case study, 171–177
benefits to industry, 140
in developing nations, 63
in electronics industry, 209, 212–213
goals of, 139, 204, 208
implementation, 139–140, 141–146
in international environmental law, 114
in life cycle analysis, 141, 201
materials flows in, 138–139
matrix system, 141, 142–144, 147
pollution prevention in, 98
product destination and, 171–172
system testing, 146–147
as systems approach, 140–141
vs. pollution prevention, 140
Design for X, 11, 139, 171, 204–205
Developing nations
Design for Environment practices in, 63
energy use in, 46, 49–50
in global energy system evolution, 57, 62–63
technology transfer agreements, 115
transition to energy sustainability, 8, 40
Dissipative loss, 31
Draft Ministerial Declaration for the Second World Climate Conference, 114

E

Earth, evolution as system, 27–28
Economic Summit of Industrialized Nations (1990), 114
Economic theory
case studies in modeling of, 64–65
intergenerational equity in, 91–92
motivation for pollution prevention, 100–107
role of, 61–62
safe minimum standard in, 93–97
sustainability accounts in, 234
utilization-oriented economy, 181–190
valuation in industrial vs. service economy, 178
Educational system
engineering curriculum, 225–226
in industrial ecosystem evolution, 16–17
role in industrial ecology, 230, 237–239
Electrification, 48–49, 55
Electronics industry
chlorofluorocarbons in, 210–211
chromated coatings in, 209–210
Design for Environment in, 209, 212–213
structure, 208
technological development in, 208–209
Emergency Planning and Community Right to Know Act, 117
Energy, Department of, 129

Energy flows
in assessing industrial evolution, 6
in closed cycle of materials flows, 26
consumption in industrialized nations, 44–45, 46
fossil fuel, 38
in global economics, 46–47
global system, rational evolution of, 56–57
in industrial metabolism model, 23–25
price shocks, 45–46
primary consumption vs. productivity of consumption, 44–45
productivity trends, 45–47
social opposition to energy consumption, 47–48
transition to sustainability, 6–8, 17–18
Energy technology trends, 48–50
Environmental Protection Agency, 5, 99, 100, 111, 149, 211
regulatory approach, 104, 105–106, 129, 194
Equilibrium models, economic, 61–62

F

Federal Facility Compliance Act, 155
Federal Insecticide, Fungicide, and Rodenticide Act, 105
Fossil fuels
benefit-pollution comparison, 39–40, 43
consumption in developing nations, 46
dissipative materials flows, 40–43
estimated current consumption, 54
estimated future consumption, 54–55
estimated supply, 50–53, 55
historical social benefits in use of, 43–44
policy questions, 38–39
social opposition to use of, 47–48
terminology, 38
in transition to sustainability, 6–8, 8–39, 55–57
U.S. consumption, 41, 54
use in developing nations, 40
Franklin, Benjamin, 220

G

General Agreement on Tariffs and Trade, 118–119
General Services Administration, 156
Germany, 109, 115, 127, 168, 183, 206, 207
Global warming, 48, 49, 56, 57 n.2
international agreements, 114
Greenhouse gases, 8, 37 n.4, 40–42, 43, 47, 49, 55–56, 57 n.2

H

Hazardous waste
data collection, 70–72
disposal cost, 149, 154–155
lead dross as, 5, 211–212
military, 150–151, 151–152, 154–155
Hydrogen, as energy source, 48, 56

I

Industrial ecology
analytical needs, 233–235
in automobile industry, 170
biological metaphor, 36–37 n.1, 130–131 n.2
definition, 130–131 n.2, 229
economic case studies, 64–65
economic growth requirements and, 90, 91
economic theory for, 61–62, 63–64
government structure and implementation of, 129–130
implementation, 125–126, 138, 230
implications for private sector, 201–207
information needs, 233
international environmental law and, 109–110
metasystem model, 231–233
principles of, 137–138
research topics, 235–237
role of university in, 16–17, 230, 237–239
social barriers to, 124–125
social context of, 9–11

sustainable development and, 5–6
as systems approach, 3–6, 17–18, 108, 124
technological development in, 229
theoretical framework, 229–230
Industrial ecosystem
assessing materials flows in, 9
barriers to evolution of, 4–5, 18, 124–125, 205–206
biological metaphor, 2, 3, 23, 25
current assessment, 8, 205–206
engineering profession in, 226
evolutionary stages, 6, 8
incentives for corporate participation in improving, 206–207
in natural ecosystem, 28, 123
primary energy consumption vs. productivity in assessing, 44
system boundaries, 1–2
Industrial metabolism, 229
concept, 23–25
energy flows in, 26
materials flows in, 25–28
measures of, 31–35
policy implications of, as holistic perspective, 35–36
research needs, 236
role of, 218
system boundaries, 25
Information needs
Air Force Pollution Prevention Program, 163
for assessment of system sustainability, 34–35
chemical reaction engineering, 224–225
defining environmental preferability, 14
in Design for Environment process, 142–143
environmental accounting, 192, 193, 200
environmental monitoring, 235, 237
environmental policymaking, 130
industrial ecology, 233, 235–237
management information and control systems, 196–197
in materials/processes comparisons, 171
materials recovery in industrial waste flows, 4–5, 18
resource substitutability, 96, 97
technological decision-making, 231–233
university-level research, 17, 233–237
waste streams, 80
Input-output analysis
data sources, 61, 62
role of, 61, 63–64, 65, 233–234
Intergenerational equity, 91–92, 146–147
Intergovernmental Panel on Climate Change, 41, 57 n.1 n.2
International environmental law
building consensus for, 111–112
command and control approach in, 110–111
eco-labeling in, 116–117
ecosystems approach in, 116
enforcement mechanisms, 119
environmental assessments in, 117–118
General Agreement on Tariffs and Trade, 118–119
impediments to systems approach in, 118–120
incentives in, 115
industrial ecology and, 109–110
internalizing environmental costs in, 112–113
market-based approaches in, 110–111
on pollution prevention, 114
precautionary principle in, 113–114
recycling and reuse in, 114–115
technology transfer in, 115–116
International implications, 8, 10, 39, 40
energy consumption, 46
resource substitutability, 92–93
U.S. environmental management, 226

J

Japan, 207
energy consumption, 44, 45, 46, 54
environmental governance in, 129–130

L

Landfill operations, 35, 37 n.5
automobile recycling residue in, 166–167
trends, 158–159

Lead
dross recycling, 5, 211–212
in gasoline, 34
in industrial waste stream, 73–78
in materials flow model, 9
solder alternatives in circuit boards, 146
Legal issues. *See also* International environmental law; Regulatory action
antitrust law, 5, 103, 131 n.12
consumer protection laws, 5
role of law in pollution prevention, 108–109, 120
Life cycle analysis, 13–15
in AT&T telephone design, 171–172
automobile recycling and, 169
implementation, 201–202
materials flows in industrial metabolism, 25–28
research needs, 236
role of, 141, 205, 234
London Declaration of Second North Sea Conference, 114
London Dumping Convention, 116

M

Market forces
in environmental regulation, 9, 10, 36, 47
global energy economies, 46–47
in industrial ecology models, 235
in industrial metabolism model, 23–25
in international environmental law, 110–111
intra-industry cooperation, 4–5, 12, 102–103, 127
materials recovery and, 4–5, 18
paper recycling, 4
reuse vs. recycling, 181–186
selling clean technologies, 226
in social cost vs. resource substitutability model, 94–95, 96
in utilization-oriented economy, 15–16, 128–129, 181–190
in voluntary pollution prevention, 100–104, 108
MARPOL Convention, 116
Material productivity, 9, 34
Materials flows. *See also* Recycling; Waste flows
anthropogenic nutrient fluxes, 28, 29*t*
in assessing industrial evolution, 6, 8
in assessing sustainability, 31–35
assessment in systems, 9
in automobile manufacturing/recycling, 165–169
in chemical engineering design paradigms, 218–220, 223–225
closed vs. open systems, 25–27
differentiating products in, 138–139
in fossil fuel use, 40–43
four-box model, 26–27
in industrial metabolism, 23–28, 31–35
natural vs. anthropogenic, 123
in sustainable development, 31
in transition to sustainable development, 15–16, 17–18
types of materials in industrial systems, 31–32
zero discharge, 8
Maximum achievable control technology, 105
Mercury, 130
Metal(s)
anthropogenic production, 123
in assessing materials flows, 9
in assessing system sustainability, 31, 34
atmospheric emissions of trace metals, 28, 30*t*
automobile recyclability, 165–167
emissions in fossil fuel consumption, 42, 54, 55
waste flow data, 72–78, 80–88
waste stream concentrations in recyclability, 78–80
Methane, 123
fossil fuel emissions, 37 n.4, 42
Military hardware
design specifications, 13, 140
green weaponry, 14, 140, 150–151, 159–163
hazardous waste generation and, 150–151, 151–152

Montreal Protocol on Substances That Deplete the Ozone Layer, 114, 115, 149, 159, 210
Municipal solid waste
annual U.S. production, 69
automobile recycling residue as percentage of, 166–167
metals in, 74

N

National Hazardous Waste Survey, 70, 71, 72, 80
National Pollutant Discharge Elimination System, 106
Natural ecosystem
evolution of Earth as system, 27–28
industrial ecology and, 2, 3, 123
industrial organization as biological organism, 25
Nitrogen oxides, 41, 43, 54
economic modeling of future emissions, 67*t*
Nuclear energy technologies, 48–49, 54, 56

O

Oil crises, 45, 46, 47, 50
Old growth forests, 95
OPEC, 45
Open-system material flow, 25–27
Organization for Economic Cooperation and Development
energy use in, 44, 45, 50
environmental accounting, 113
Our Common Future, 65, 90, 228
Ozone-depleting chemicals, military use of, 149, 152, 160

P

Paper/paper products
economic viability of recycling, 4
regulatory control, 105–106
vs. reusable products, 1
Pesticides, 128
Plastics, in automobile recycling, 168
Pollution prevention. *See also* Air Force Pollution Prevention Program
atmospheric emissions of trace metals, 28, 30*t*
chemical engineering design paradigm, 223–225
conceptual development, 98, 137, 201, 222–225
economics of voluntary compliance, 100–104, 108
education/training courses, 153–154, 225–226
engineering design in, 223, 225–226
future needs, 63
industry benefits, 98–99
international law mechanisms, 110–118
intra-industry cooperation, 102–103, 127
as market value, 10
by medium, 35, 105
nitrogen oxides in, 43
regulatory solutions, 107
risk assessment and, 99
role of law in, 108–109, 120
sulfur oxides in, 43
in systems approach, 17–18, 35–36, 137
taxation incentives for, 10–11
via enforcement, 106–107
via permitting, 106–107
vs. historical benefits of fossil fuels, 39–40, 43–44
Pollution Prevention Act, 105, 109
Postconsumer waste, 69, 70–72
Private sector. *See also* Accounting practice
benefits of Design for Environment, 140
benefits of pollution prevention for, 98–99
economic motivation for pollution prevention, 100–104
environmental leadership by management, 203
in evolution of industrial ecosystems, 11–13
industrial organization as biological organism, 25
industrial technology and, 201–207

in international environmental law development, 111–112
intra-industry cooperation, 4–5, 12, 102–103, 127
life cycle assessments in economic decisions, 13–15, 141
motivation of, research needs on, 236–237
organizational structure of firms, 37 n.8
total environmental cost accounting, 15
in utilization-oriented economy, 15–16, 128–129, 181–183
voluntary initiatives vs. government regulation, 47, 100
Product design/development. *See also* Design for Environment
automobile recyclability, 167–170
consumer interest in green products, 140
customer specifications in, 13, 140
engineering education and, 16–17, 225–226
engineering profession in, 217–218
environmental assessment methodology, 172–176
environmental factors in, 11, 12–13
geographic impacts, 146
government intervention, 125
in industrial ecology concept, 126
intergenerational considerations, 146–147
life cycle assessment in, 13–15, 141
participants in, 205
product complexity/materials and, 138–139
supplier management systems in, 12–13, 142, 155–156
telephone, case study, 171–177
total environmental cost in, 15, 125–126
in utilization-oriented economy, 188
Product labeling, 116–117
Productivity of materials, 9, 34

R

Reaction products, 222
Recycling
in Air Force Pollution Prevention Program, 159
in assessing industrial evolution, 6, 8
in assessing system sustainability, 31, 34
assessment of environmental cost in, 172–173
in automobile industry, 165–167
closed loop model, 179–183
complexity of product design and, 138–139
concentration in waste stream and, 78–80, 88
as conclusion of materials flow, 31
information needs, 4, 18
international agreements, 114–115
of lead waste, 73, 74–78, 211–212
liability concept in, 181–186
market force barriers to, 4–5, 18
optimizing use of goods vs., 183–186, 189–190
parts labeling in manufacturing process, 167, 174
plastics, 168
regulatory barriers to, 5, 18
remanufacturing, 129, 132 n.16, 166
take-back regulations, 127–128, 129, 139, 168, 183, 206, 207
telephone, 176
waste flows in systems approach, 3–4, 17–18
Regulatory action
automobile industry and, 169–170
chlorofluorocarbons in electronics industry, 210–211
command-and-control approach, 104, 109, 110–111, 126, 138, 212, 213
in corporate accounting, 194
economics of voluntary pollution prevention and, 101–104, 108
encouraging use vs. production of goods, 128–129
federal approaches, 104–107, 129
government purchasing procedure as, 155–158
hazardous classification of lead dross, 5, 211–212
in holistic perspective, 35–36
indications for, 10, 47, 103–104
in industrial ecology models, 235

in industrial metabolism perspective, 35–36
market orientation of, 9, 10, 125–126
as obstacle to industrial ecosystem evolution, 5, 10, 18, 47, 209, 212–213
on packaging, 206
product complexity and, 139
product design process, 125
in promoting intra-industry cooperation, 127
prospects, 107
in social cost vs. resource substitutability model, 94–95
state level, 104
structure of government and, 129–130
systems approach in, 126–128, 129–130
take-back regulations for industry, 127–128, 129, 139, 168, 206, 207
technical knowledge in, 212–213
Resource Conservation and Recovery Act, 5, 71, 106, 131 n.11, 211
Rio Declaration, 113

S

Safe minimum standard, 93–97
Sherwood diagram, 69–70, 78, 88
Service economy, valuation in, 178
Social values
barriers to ecological systems perspective, 124–125
ecological-economic linkages needed in, 124
in industrial ecology, 9–11
opposition to energy consumption, 47–48
pollution vs., in fossil fuel use, 39–40, 43–44
resource substitutability and, in limiting scenarios, 10, 93–97
technological development and, 220–222
Solar energy, 48, 49
Steady-state systems, 26. *See also* Sustainable systems
Structural economics, 61, 62, 64
Substitutability of resources, 37 n.7, 91, 92–93
social costs and, in limiting scenarios, 10, 93–97
Sulfur
anthropogenic emissions, 41–42
waste flow of, 32–33
Sulfur oxides, 41, 43, 54
economic modeling of future emissions, 66
Summit of the Arch, 114
Superfund Amendments and Reauthorization Act, 71
Sustainability
economic modeling, 234
industrial ecology and, 5–6
intergenerational equity calculations, 91–92, 146–147
materials flows in assessment of, 31–35
meaning of, 5, 90–91, 228
research needs, 236
resource substitution for, 90–91, 92–93
technological change for, 228–229
transition to, 6–8, 15–16, 17–18, 38–39, 55–57, 237–239

T

Take-back regulations, 127–128, 129, 139, 168, 183, 206, 207
Taxation
to encourage evolution of industrial ecosystems, 10–11, 126
energy use and, 45–46, 47
Taylor, Frederick, 37 n.8
Technological development
attitudes toward, 220–222
research needs, 237
role of, 228–229
Telephone design, case study of, 171–177
Thoreau, Henry David, 220–221
Total quality management, 11, 12, 205, 211
Toxic Release Inventory, 70, 71
Toxic Substances Control Act, 105
Treaty on European Union, 114

U

United Nations Conference on Environment and Development, 62, 64, 99, 113–114, 228
Utilization-oriented economy, 181–190

V

Volatile organic compounds, 41, 42, 132 n.18

W

Waste flows. *See also* Materials flows
of chlorofluorocarbons, 33
concentration in, and recyclability of metals, 78–80, 88
current estimates, 69
data sources, 70–71, 72, 80
estimating resource values in, 72
fast food industry, 131 n. 10
industrial, 40–41, 71–72
lead in, 9
metals in, 73–78
in military settings, 152, 154–155
recoverable materials in, 69
recycle/reuse loops, 179–183
of sulfur, 32
in systems approach, 3–4
waste reduction strategies, 178–179
Waste sinks, 2
Wetlands, 95
World Charter for Nature, 114
World Commission on Environment and Development, 90, 228
World Energy Council, 41

Z

Zero discharge, 8, 100, 223